CLICK AND KIN

T0324525

Transnational Identity and Quick Media

Edited by May Friedman and Silvia Schultermandl

Click and Kin is an interdisciplinary examination of how our increasingly mobile and networked age is changing the experience of kinship and connection. Focusing on how identity formation is affected by quick media such as instant messaging, video chat, and social networks, the contributors to this collection use ethnographic and textual analyses, as well as autobiographical approaches, to demonstrate the ways in which the ability to communicate across national boundaries is transforming how we grow together and apart as families, communities, and nations.

The essays in *Click and Kin* span the globe, examining transnational connections that touch in the United States, Canada, Mexico, India, Pakistan, and elsewhere. Together, they offer a unique reflection on the intersection of new media, identity politics, and kinship in the twenty-first century.

MAY FRIEDMAN is an associate professor in the School of Social Work at Ryerson University and author of the award-winning *Mommyblogs and the Changing Face of Motherhood*.

SILVIA SCHULTERMANDL is an assistant professor in the Department of American Studies at the University of Graz in Austria.

Click and Kin

Transnational Identity and Quick Media

EDITED BY MAY FRIEDMAN AND
SILVIA SCHULTERMANDL

UNIVERSITY OF TORONTO PRESS
Toronto Buffalo London

ISBN 978-1-4875-0000-9 (cloth)   ISBN 978-1-4875-1996-4 (paper)

Library and Archives Canada Cataloguing in Publication

Click and kin : transnational identity and quick media / edited by May Friedman and Silvia Schultermandl.

Includes bibliographical references and index.
ISBN 978-1-4875-0000-9 (cloth).   ISBN 978-1-4875-1996-4 (paper)

1. Transborder ethnic groups – Social conditions. 2. Transnationalism – Social aspects. 3. Kinship – Social aspects. 4. Social media. 5. Families. 6. Identity politics. 7. Feminist theory. I. Friedman, May, 1975–, author, editor II. Schultermandl, Silvia, 1977–, author, editor

HM1271.C55 2016   305.8   C2015-907811-3

This book has been published with help from grants from the University of Graz. The authors also gratefully acknowledge the support of the Faculty of Community Services at Ryerson University.

University of Toronto Press acknowledges the financial assistance to its publishing program of the Canada Council for the Arts and the Ontario Arts Council, an agency of the Ontario government.

**Canada Council   Conseil des Arts
for the Arts   du Canada**

ONTARIO ARTS COUNCIL
CONSEIL DES ARTS DE L'ONTARIO

an Ontario government agency
un organisme du gouvernement de l'Ontario

Funded by the   Financé par le
Government   gouvernement
of Canada   du Canada

To our children, Noah, Molly, Aviva, Isaac, Kira, and Sasha, for constantly expanding our understanding of family and new media. OMG, you make us LOL.

Contents

Acknowledgments

This book is the culmination of many webs of support, both digital and corporeal. We are so very appreciative of this robust and dynamic collaboration and are grateful to everyone who helped this volume come to completion.

We especially thank the contributors to this volume, who ensured that conversations about kinship, transnationality, and quick media were engaging, compelling, and full of surprises. The collection would not have been possible without the support of the University of Toronto Press. Likewise, we give thanks for support from the Faculty of Community Services at Ryerson University and the Faculty of the Humanities at the University of Graz.

Most of all, we are incredibly grateful to our families for conveying the urgency of studying transnational kinship and digital connections day in and day out.

CLICK AND KIN

Transnational Identity and Quick Media

Introduction

MAY FRIEDMAN AND SILVIA SCHULTERMANDL

When we don't know an answer, we often look it up online. But what are we to do when we, ourselves, become the question? As our lives become both increasingly geographically diffuse and overwhelmingly networked by communication technologies, existing ideas about families, identities, nationalities, and bodies may provide insufficient modes of analysis. Understanding the tensions and congruities among ourselves, our communities, and the world around us may become increasingly uncomfortable; the notion of stable identity may feel like a vanishing ideal. The bridges and the fissures between connections and identities, between humans and machines, between family and "kin" have led us to this anthology.

Click and Kin aims to consider the implications of an era of rapidly increasing transnationalism and multimedia exposure as a means of negotiating kinship and connection. In particular, this volume explores the interstices between coherent national and cultural identities and examines the ways that technology – cell phones, Skype, email, Facebook, Twitter, SMS, and others – simultaneously disembodies and re-embodies our experiences of connection over distance, with implications for our singular and collective identity formation. These forms of disembodied and re-embodied practices of identity negotiation raise questions about how we inhabit and interact through time and space; how we rethink identity and our ideas of close and far relations, in terms both of kinship and of physical distance; how the emergence of new media technologies generates new perspectives on bifurcated and hybridized lives; how and where we find our "chosen family"; and how the family itself is undefined and redefined in the combination of mass mobility and cheap and accessible international interaction. These questions are prompted by an epistemology of wonder, a means of remaining open to and inspired by liminality and confusion. We aim to explore a transnational sensibility which honours the ambiguity of borders, families, and identities and views "a lack of fixity as simultaneously inevitable and rich in possibility" (Friedman and Schultermandl 5). The turbulence and reimagination in relationships and subjectivities which informs this transnational sensibility challenges prevalent notions of identity and kinship, particularly when those notions are constructed and maintained through quick media.

Whether it's loved or loathed, our lives are increasingly governed by "quick media," through our professional experiences, in our relationship with popular culture, and with our families far and near. Quick media is an umbrella term for the cheap, easily accessible, and omnipresent tools of communication which allow us to connect to each other spontaneously and effortlessly and which include both messaging platforms such as text and Skype and social media outlets such as Facebook and Twitter. Given the availability of these tools and a certain level of media literacy, especially among millennials, quick media constantly accompany us through our daily routines, whether we use them ourselves or are privy to their excessive use around us. Quick media are a fairly recent phenomenon but one whose impact we see in the fact that human engagement with media has increased exponentially. We Skype, text, and tweet to a degree that we didn't – and couldn't – only a decade ago, especially before the "2.0-ization" of network technologies, when these semantic neologisms weren't part of our vocabulary, much less of our sense of self. Unlike personal websites and emails, quick media have less elaborate interfaces, are more portable and transferable, and above all, work on the basis of two people using the same thing: I can email from Yahoo to Gmail, but when I use Twitter I am confined within the format of all Twitter accounts.

This new use of media affects our sense of self. While in the past media engagement was a more passive and receptive undertaking (as in the examples of television, film, and publishing; Shohat and Stam), we are active participants in an era of networked connections. Never before has our engagement with media been so dialogic, so varied, and so easily accessed by non-professionals. As a result, our relationships and identities are mediated through the tools we use – our computers, tablets, and cell phones and the software which governs them. Beyond national and international states of being, this volume aims to expand our understanding of kin, considering the ways that new technologies allow for a broadening of kinship that may mitigate the limitations of physical geography, allowing instead for increased connection on the basis of affinity (Karlsson). Quick media give us the feeling that we are always present, within reach, and connected to the world. Anybody who has ever texted after a flight's arrival knows the feelings of empowerment and relief facilitated by quick media technologies. At the same time, anybody who has ever waited to receive such a text message knows the feelings of disempowerment and anxiety prompted by a loved one's lack of availability. That is how much we frame our daily lived experiences through our online presence.

Who We Are, Where We Are From, Who We Are Becoming ...

MAY: How different would this book, and our relationship, be without quick media? Would we have maintained the same intimacy without our capacity to use email and Skype to stay on top of one another despite the fact that, because of our different time zones, I drop my kids off just as you pick yours up, leaving us very little child-free time in the middle to get anything done?

SILVIA: I can think of one thing that would be different – I would not be seeing your kids pop up once in a while on the screen when they check in on you, or you mine.

MAY: I wouldn't know what your home looks like. You would never "see" my office.

MAY: I don't think it's a stretch to say that our robust and dynamic relationship is largely dependent on quick media – that it might have withered without it. We've seen each other – what – four times in person?

SILVIA: That's ridiculous, right? But the interesting thing is that there are benefits to both our scholarship and our relationship from us being far apart. I revel in our sameness and our differences, the ways we are broadly doing the same things (launching academic careers while raising young families) but with extraordinary differences.

MAY: Yes, the specific circumstances in which we work also shape our collaboration. Like – you go to France for the weekend. That mystifies me.

SILVIA: Ha! But also how our universities run differently, the different childcare options we have access to ...

MAY: And also the minutiae of our lives are different anyway, independent of nation – the ages of our children, the differences in our neighbourhoods, our different disciplinary orientations. I feel like quick media technology amplifies both the sameness and the difference.

SILVIA: But also offers strategies to work with sameness and difference creatively and productively.

MAY: Perhaps that is what we love about this collaboration and also about quick media – both give us a means of exploring the details more thoroughly – not necessarily to come to any conclusion, but just to revel in a thorough mapping of our lives and thoughts, separately and together.

MAY: It allows us to be in a constant conversation, despite our distance and *about* our distance.

Kinship and the Imagined Family

We are thinking of family in the sense of an *imagined family*, adopting Benedict Anderson's understanding of the nation-state as an imagined community. Anderson argues that "all communities larger than primordial villages of face-to-face contact (and perhaps even these) are imagined communities" (6). This is also true for the kinship constellations attained through quick media technologies. In their most optimistic sense, these imagined families strive for a feeling of universality, gained precisely by the equal access to quick media technologies as a means of performing identities, building communities, and transcending social boundaries. This idealistic and potentially empowering aspect of the imagined family goes back to Marshall McLuhan's idea of the global village, a sense of community which can be established in accelerated and uncomplicated ways because media can cross the time-space distinction that roots us in our particular locations (31). As celebrated instruments with the potential of making our lives more connected in an age of hypermobility, quick media technologies allow us to imagine communities and families differently.

But the very notion of a global village is shot through with contradictions. Do we truly know another when we connect online? How are our practices of online communication, bound by their own constraints and rituals, yet another venue in which we perform acts of de-realization, by which we temporarily become the subject of our virtual interfaces but not ourselves? As the dialogue we just shared shows, our strong kinship ties depend on both our embodied and disembodied encounters, but we realize that this is if not unique then certainly special given the special context of feminist scholarship in which we met. The fact that we ourselves "leaked" this conversation about kinship and academia shows our agency as scholars who can productively use personal context to solidify abstract ideas about kinship, a much different form of public dissemination of personal ideas than prevalent forms of cyber-control.

Furthermore, this shift towards networked connection has had differential effects globally. On the one hand, the digital divide has ensured that networked technologies have not been evenly distributed (Mehra, Merkel, and Bishop 782). The physical infrastructures required for Internet access, for example, are sporadically available in rural areas, even within Western countries. Likewise, many parts of the developing world lack the practical requirements for online access. Just as we begin to understand the impact of the digital divide on global communication, however, we see that impact lessened by an increased reliance on cell phone technologies, which are much more readily and affordably available in places where computers may be scarce.

Beyond issues of access, there is the consideration that just because technical facilities enable us to connect with someone far away does not mean that we can automatically relate to and empathize with this person. The sense of familiarity implied in the idea of a global village presumes that there is a sense of kinship based on the belief that we are all alike because we can all access the same virtual world. Gayatri Chakravorty Spivak has offered an astute critique of this notion that a global village can be built through communication networks. The notion of a global village fails to acknowledge the innate difference in power and privilege between different global societies despite the fact that their various members might be equally literate in, and engaging with, online technologies. The idea of the village, so Spivak contends, is supposed to produce a feeling of connectedness with remote cultures, which is not sustained by the same lived experiences or fails to account for inherent asymmetries. To overlook the contradiction implied in the words "global" and "village" is a very sanguine attempt to force similarities where there are none. This unilateral expansion of the idea of kinship onto "other" cultures is, according to Spivak, "colonialism's newest trick" (330).

These Janus-faced characteristics of quick media themselves amplify the contradictions of the postmodern self. Fundamentally, it is futile to attempt to tally the radical and oppressive potential of quick media because of their pervasiveness. Transnational communication and identity has become virtually indiscernible from the means of that communication, suggesting that a reckoning with the myriad implications of quick media is both fruitful and timely. The question is not so much whether quick media technologies have the potential to reinforce or disrupt traditional family structures by allowing us to build and maintain kinship ties: the question is how their impact on the formation of kinship ties is *different* because of the mediated nature of our interactions. Benedict Anderson argues about the nation-state that communities are not to be distinguished as false or genuine, "but by the style in which they are imagined" (6). Quick media is the location of those communication styles which have drastically altered our means of imagining community by selectively fostering kinship ties, based not only on common genealogies but also on common values.

In postmodern identity theories, especially in relation to processes of hybridization, a similar rhetoric of being and becoming has revolutionized ideas about the self in transition. Stuart Hall's use of "being" and "becoming" as concepts describing bifurcated identities acknowledges the "critical points of deep and significant *difference* which constitute 'what we really are'; or rather – since history has intervened – 'what we have become'" (225, emphasis in original). In this sense, Hall proposes an understanding of cultural identity which is a "matter of 'becoming' as well as 'being.' It belongs to the future as much as to the past. It is not something which already exists, transcending place, time, history, and culture" (225). The logical conclusion from Hall's distinction between being and becoming is that cultural identities are "unstable points of identification" (226) which strive on the inherent contradictions between the multiple forms of selfhood resulting from the multitude of forces acting upon the self. To extend this analysis to the realm of kinship is to consider family as a thing made, rather than a static marker. Working with and through transnational identity and quick media, this collection aims to take up kinship as an unfolding series of dialogues, instead of a taken-for-granted and narrowly defined focus on family.

Kinship through Quick Media

In a postmodern and globalized age, the notion of kinship is intertwined with ideas about the self-in-relation and of place as a node of multiple and contradictory cultural currents. By extension, the family itself becomes dialogic and in flux. Notions of kinship mirror these changes to traditional concepts of family; quick media technologies, in particular, open up a breadth of new places where we look for and find our kin and where we reconsider who figures in our lives as kin. If kinship, according to Linda Stone, is "an ideology of human relationships" which brings into sharper focus the "cultural ideas about how humans are created and the nature and meaning of their biological and moral connections with others" (6), quick media technologies can facilitate new, faster, and more stratified interactions in a seemingly democratic virtual space. By their very nature, quick media foreground the personal and subjective and are part of a long tradition of self-referential writing with the intention of making these expressions of the self available to others.[1] This provides alternative means of identification and kinship ties to embodied lived experiences of human interaction. The kinship ties we create and sustain online challenge ideas about the monolithic family and disrupt the family as an institution.

Families of origin provide a robust site of analysis for identity formation. Drawing on the work of feminist motherhood scholars (O'Reilly; Hays; Kinser), an exploration of family may expose the nuances of identity versus practice (Rich). Motherhood as identity is rooted in the expected ideals of motherhood and takes place within institutions of patriarchy. By contrast, motherhood as practice takes up the multiple, contradictory, and interrupted moments of both caring labour and emotional work that compose parenting practice. Rich contends that the institution of motherhood (motherhood as identity) cannot be divorced from conditions of patriarchy. By contrast, an examination of motherhood practices – by people of all genders – allows a much richer and more divergent picture to emerge.

Feminist analyses of motherhood seek to unpack the ways that both "mother" and "family" have often been reduced to normative concepts that ignore or, worse, repress difference. Looking far, far beyond the realm of Beaver Cleaver and the mythical white, middle-class nuclear family living in one home (and one nation), with a shared, coherent identity, kinship must be understood as a much more complicated and diffuse motif. A recurring trope within media and politics, especially in

---

1 In *The Veil and the Mirror: An Overview of American Online Diaries and Blogs*, Viviane Serfaty argues that self-representational writing has a long-standing history in American literature, but that with the new availability of online platforms as outlets for these forms of life writing, it offers new insights in the autobiographical practices of men and women alike.

the United States, is the notion of the nuclear family as "in trouble." Far from seeing this as a problem, we seek to expose and explore the ways that this "trouble" is a site of tremendous opportunity. A "troubled" version of family may cast light on bifurcated families, transnational families, and families who do not neatly ally with popular representations of kin. Instead, drawing on work from queer and transgender theorists (Butler; Halberstam) and transnational scholars (Shohat; Hegde; Kaplan, Alarcón, and Moallem) allows for families to be contested, inconsistent, and self-defined. For example, Hosu Kim's analysis of Korean birth mothers' online writing about their children who have been adopted disrupts static notions of family by presenting motherhood independent of the traditional caregiving tasks which often define mothers. Memoirs by trans folks who have transitioned while parenting similarly disrupt the perceived intractability of "mother" and "father" as organizing categories (Bornstein; Ladin; Finney Boylan).

Quick media have tremendous potential with respect to the self-definition of kinship. As the networked potential of global communication grows, people are increasingly seeking chosen family through online cultures. These "families" suggest that kinship may be something we do rather than who we are, pointing to a dialogic and agentic reckoning of family. For hybrid and transnational subjects who live outside of the mythic norm, finding similar subjects online may lead to nourishing interactions that are truly familial in scope (Karlsson). Parents of children with disabilities, for example, may make stronger connections with allies in common than with their families of origin; likewise, queer folks may find stronger family ties "in the computer" than in their homes.

The combinations of both the limits and increases to access have had important implications for families who have been separated by immigration, violence, or other forms of displacement. For example, Mirca Madianou and Daniel Miller document the implications of new technologies for Filipina women who are working as caregivers in the United Kingdom. They discuss the ways that "polymedia"[2] allow for a

---

2 Madianou and Miller define polymedia by pointing out the following criteria: "the first condition for the emergence of polymedia is the availability of a plethora of media choices, when a user has at least half-a-dozen media at their disposal that their household can readily afford. Secondly, users must have the skills and confidence to use these digital media. The third condition is that the main costs are infrastructural, such as paying for the hardware and the Internet connection cost, rather than the cost of individual acts of communications" (137).

new participation in family life back home. Yet the economic disparities that result in many Filipina mothers being separated from their
children are not easily addressed despite the implications of that separation being mitigated by quick media. Furthermore, the effects of
racism and colonialism which lead to these economic disparities may
likewise have effects in online interactions. Lisa Nakamura contends
that in the IT industry, prevalent racial stereotypes are being reproduced by means of creating cybertypes which adhere to essentialist and
racist depictions of foreign and domestic minorities. Drawing from a
variety of approaches from critical race theory (e.g., bell hooks, Audre
Lorde, and Vijay Prashad), Nakamura extends her analysis of the politics which describe cybertypes discursively both in the online and the
offline spaces of Internet technology. Existing discursive structures may
be amplified in online interactions, suggesting that "the cultures and
practices developed around media forms provide an analytical space
from which to examine how the global is performed, reproduced and
contested within the material specificities of everyday life" (Hegde 6).

While communication technologies may amplify existing oppressive structures, they may also provide spaces through which to interrupt existing tropes. Shelley Park suggests that hegemonic parenting
practices may be transformed by the use of technologies, resulting in
"cyborg mothering," defined as motherhood mediated through technologies of parenting (baby monitors, educational videos, and other
paraphernalia). Through the interaction of human and machine, Park
notes the capacity for technologies of co-presence, which "allow us to
be present to and with others in ways not tied to the physical facticity
of the body. These technologies thus permit us ... to inhabit space and
time differently – in ways that enable resistance to bourgeois family
norms" (63). Friedman takes the notion of the cyborg mother further,
considering the ways that mommy blogs allow mothers to undertake
parenting as a dialogic enterprise, writing their stories and questions to
and with one another and informed as parents by a chorus of support
and dissent (*Mommyblogs*). This is not to say that quick media deserve
ceaseless veneration. The constraints and uneven access to quick media
technologies – their role in extending surveillance, reproducing power
relations, and generating new modes of exclusion – also mirror and
in part amplify prevalent social inequalities, especially in a globalized
context. If gender inequality, as Judith Lorber has pointed out, "is
not an individual matter but is deeply engrained in the structure of
societies" (7), these structures are not always easily overcome once we

go online. But what happens is that these structures are engrained in new and emerging patterns which only become accessible once we use quick media technologies. This in itself allows for new possibilities of interaction that respond to existing limitations.

Quick Media as Mediation of Transnationalism

The new kinship ties we foster through our use of quick media technologies can best be understood as a dynamic set of practices and exchanges which affect literal and metaphoric forms of border crossings and transnationalism. Similar to the proliferation of quick media technologies, this idea of the transnational is connected to contemporary phenomena of globalization and hypermobility, but it is not synonymous with the global per se.[3] The transnational is a tool to get at the heart of preconceived notions of identity and culture,[4] akin to the conceptual tool of gender as a means of dismantling prevalent notions of masculinity and femininity. As Briggs, McCormick, and Way note, "'transnationalism' can do to the nation what gender did for sexed bodies: provide the conceptual acid that denaturalizes all their deployments, compelling us to acknowledge that the nation, like sex, is a thing contested, interrupted, and always shot through with contradiction" (627). In this vein, the transnational destabilizes concepts of identity as solid and complete, and highlights the fluid and ever-changing nature

---

3 A useful distinction between the global and the transnational comes from Francoise Lionnet and Shu-mei Shih's introduction to their collection of essays, *Minor Transnationalism*: "Whereas the global, in our understanding, is defined vis-à-vis a homogenous and dominant set of criteria, the transnational designates spaces and practices acted upon by border-crossing agents, be they dominant or marginal. The logic of globalization is centripetal and centrifugal at the same time and assumes a universal core or norm, which spreads out across the world while pulling into its vortex other forms of culture to be tested by its norm. It produces a hierarchy of subjects between the so-called universal and particular, with all the attendant problems of Eurocentric universalism. The transnational, on the contrary, can be conceived as a space of exchange and participation wherever processes of hybridization occur and where it is still possible for cultures to be produced and performed without necessary mediation by the center" (5).

4 We are hence interested in what Sanjeev Khagram and Peggy Levitt call "philosophical transnationalism," i.e., one which is "based on the metaphysical view that social life is transnational to begin with" (8). The inherently dialectical approach of transnational studies is also at the centre of Laura Doyle's review of phenomena of intersubjective and intertextual formulations of transnationalism proposed in her essay "Toward a Philosophy of Transnationalism."

of identities, both of individuals and of families. At the centre of our investigations of kinship lies our interest in what we call a transnational sensibility, which "begins by looking at life in and on borders, yet it goes beyond these inarguably contested subjects" (Friedman and Schultermandl 5). The transnational

> can be fruitfully applied to hybridized subjects as well as to those whose identities are presumed to be fixed. As such, this transnational sensibility sees a lack of fixity as simultaneously inevitable and rich in possibility. A transnational sensibility is both a methodology and a mode of inquiry: a way of seeing and deliberately *not-knowing*, a way of living within the spaces between questions and answers. (Friedman and Schultermandl 5, emphasis in original)

Since kinship ties facilitated through quick media technologies allow for new but not newly unequivocal forms of identity building, they exemplify the inherent ambivalence well-known to transnational subjects. Quick media are therefore a means not only of compressing spatial and temporal distance in a globalized world, but also of mediating a transnational sense of self in relation to one's kin, near or far. The use of quick media technologies also facilitates a sense of selfhood which strives on the inherent contradiction of being together, yet apart, a condition which locates personal identity in two opposing states of being. In *Who Sings the Nation-State? Language, Politics, Belonging*, Judith Butler and Gayatri Chakravorty Spivak theorize this notion of a dual state: *state* as a national entity with its legal boundaries and discursive paradigms versus *state* as the condition that describes the circumstances of their writing:

> So: how do we understand those sets of conditions and dispositions that account for the "state we are in" (which could, after all, be a state of mind) from the "state" we are in when and if we hold rights of citizenship or when the state functions as the provisional domicile for our work? (2)

This duality of the term "state" pertains to the questions of national belonging and to any other discursively conscripted sense of belonging to a community. Given the constructedness of these "states" within the imagined communities and families that frame our sense of self, the fissures between these states are indicative of the sense of ambivalence which lies at the basis of selfhood in a postmodern, transnational, and cybernetic era. It is a bold statement when Butler and Spivak suggest that "there can be no radical politics of change without performative contradiction" (66), but one which ultimately captures the essence of quick media's influence on kinship in a transnational and mediated context.

To be sure, this sense of ambivalence is not newly prompted by the emergence of quick media technologies. In fact, ambivalence, as Zygmunt Bauman observes, has been a driving force throughout Western modernity, but one which is unrightfully categorized as a default or disorder. Bauman considers the concept of ambivalence rich in possibility to reconceptualize social orders, and in the particular context of quick media, this convergence of transnational identity and kinship mediated through quick media adds new layers of contradictions to an already dynamic concept of the self-in-making. This convergence may also consider the ways that we live at the intersection of machine and organism. Matthew Wilson's analysis of Donna Haraway's concept of the cyborg as a figuration which articulates ideas of being and belonging dovetails with transnational concepts of identity. In Wilson's context, cyborgs go beyond identity politics (such as were prevalent in the 1990s) where essentialist ideas of selfhood described and defined discourses of political agency: "figurations transcend rationalities and involve multiplicity, but motivate a kind of objectivity through embodied perspective" (501). This idea of multiplicity resonates with the transnational as a concept of identity which emphasizes contradictions and tensions for the sake of creating a productive space of negotiating selfhood beyond mere binary constructions. In postmodern identity theories, especially in relation to processes of hybridization, a similar rhetoric of being and becoming has revolutionized ideas about the self in transition.

The Chapters

So what happens when we go online in search of ourselves? The individual chapters of this collection suggest a variety of answers to this question and, in turn, prompt new ones. By picking up the notion of a "transnational sensibility" and extending the relevance of this concept into the virtual world, all chapters centre on the ways that quick media facilitate alternative modes of identification. While all chapters pursue this notion in one way or another, they each position different sets of caveats towards the assumption that quick media technologies are utopian forms of self-inventions; the chapters equally resist a strict demonization of new technologies. This strong sense of awareness of the potentials and limitations of the different technologies shapes the individual authors' understanding of kinship. It is along these lines, the question of self-invention and kinship building in the context of different quick media technologies, that we would like to emphasize four thematic strands: first, quick media and the connections between the individual and an imagined community; second, quick media and emergent, established, and alternative views of identity; third, re/envisioning the self in cyberspace as an alternative to lived identity; and last, quick media as a substitute for transnational encounters when embodied encounters are not possible.

The chapters approach the topic of transnational identity at the intersection of kinship and quick media through a variety of lenses, articulating their insights in autoethnographic essays, critical discourse analyses, literary analyses, philosophical explorations, and anthropological case studies. By reading quick media messages – blog entries, Facebook postings, instant messages – "as rhetorical artifacts" (Gurak et al.), they collectively stitch together a cultural history of the online practices so symptomatic of our communicative behaviour since the proliferation of quick media.

Between the Individual and an Imagined Community

Several essays in the collection investigate the dynamics within digital diasporas with particular attention to the processes of identification between the individual and one or more imagined communities. As is symptomatic of digital diasporas in general, the ones discussed here "use the Internet to negotiate their identity and promote solidarity; learn, explore, and enact democratic values; and mobilize to peacefully pursue policy influence, service objectives, and economic participation in the homeland" (Brinkerhoff 2). Michel Laguerre offers a similar definition of digital diasporas but contends that the subject at hand in the identity constructions of diasporans occurs in relation not only to their homeland but to their present location as well as in relation to the trajectory in between the two (50).

The resulting ambivalent relationship with technology through both identity construction and kinship is typical for second-generation immigrants and their communication with family members "back home." Their identity negotiations in relation to this imagined community often build on existing close bonds, which are then renewed and modified through digital interactions, intensifying emotional and cultural ties that remain alive through quick media usage. This is certainly the case with Laura E. Enriquez's chapter, "'I Talk to My Family in Mexico but I Don't *Know* Them': Undocumented Young Adults Negotiate Belonging in the United States through Conversations with Mexico," an ethnographic study of undocumented and recently legalized Mexican-origin young adults and their use of quick media as a means of creating, sustaining, and negotiating their identities within the Mexican diaspora. What Enriquez shows is that quick media create the impression of closeness to kin with whom these young adults have very little embodied experience.

The strong political consciousness of digital diasporas is easily traceable through ethnographic analyses of cyberspace behaviour, including the frequency and intensity of the interpersonal exchanges and the general postings. In many cases, there is a direct link to the occurrence of political events, which prompts the outpouring of messages and data through quick media technologies. In "'Learning and Practising Democracy': Digital Diasporas and Negotiating a Transnational Civil Society," M. Tina Zarpour undertakes a qualitative analysis (with some descriptive numbers) of the effect of "digital diasporas" in the context of online message forums maintained by Iranian diasporic communities. Iranian diasporic communities, for instance, immediately responded to newly emerging nationalism in Iran by debating the question of what constitutes Iranian identity in the various homelands. Both identification with existing communities and the creation of new ones can be traced through digital encounters in the cyber world. Zarpour examines a particular site frequented by diasporic Iranians, noting the shifts in interactivity facilitated by the convergence of digitality and diaspora, which leads her to conclude that the site "provided a forum for the safe expression of a diverse range of political, ethnic, and religious difference not found in face-to-face interactions among migrant Iranians." Blogging and posting alters community dynamics, which usually manifest themselves through social gatherings of diasporans, to allow for a political engagement going well beyond what is often derogatively termed "Facebook activism."

Quick media can also provide a platform for a transnational sensibility with regard to contemporary popular cultural phenomena. An emergent networked South Asian identity that resists the master narratives of Indian and Pakistani subjectivities in a post-war era, for instance, allows users to create kinship ties beyond the India-Pakistan partition in 1947, the Indo-Pakistani War of 1971, or the ongoing militarized conflicts in the two nation-states. In light of this historically rooted antagonism between India and Pakistan, the Facebook fan site of tennis player Sania Mirza both replicates and subversively undermines prevalent nationalist discourses.

In their chapter, "Negotiating Womanhood and South Asian Nationalisms: Blurring Borders and Identities in Social Media," Aparajita De and Shekh Moinuddin offer a critical discourse analysis of the Facebook postings in which fans and critics of the tennis player challenge the totality of the nation-state both by critiquing it and by making its systematic constructedness visible. Facebook is also the main medium of communication through which kinship ties which overrule the India-Pakistan partition become possible. By exploring the interactivity of South Asian message boards frequented by Indian and Pakistani people, De and Moinuddin exemplify to what degree social media may disrupt nationalist discourses and foster a sense of kinship between politically antagonized nation-states.

Shaping Identities

Users of quick media technologies are able to negotiate between individual identities and that of an imagined community because such technologies offer increasingly creative means for participatory cultures – i.e., cultures in which "members also believe their contributions matter and feel some degree of social connection with one another (at the least, members care about others' opinions of what they have created)" (Jenkins xi). The various forms of participatory cultures, including affiliation, expression of opinions, collaborative problem-solving, and circulations (cf. Jenkins xi–xii), depend largely on the medium of dissemination.

For instance, in the case of Lesbian Web Series (scripted series released on the Internet), identity formation between the protagonists of the series and their strong, active fan base occurs through YouTube videos, which appropriate original footage to create expressions of identification. These YouTube videos are not collages of random original footage but carefully arranged expressions of a collective identity coalescing around the protagonists. Because YouTube is ultimately a transnational medium, the videos the fans post speak to this sense of a transnational kinship. In her chapter, "Queering 'Web' Families: Cultural Kinship through Lesbian Web Series," Julia Obermayr examines the construction of kinship ties across lines of transnationality, sexuality, and virtuality in the format of Lesbian Web Series, online programs about queer lives that are accessed around the world. The chapter shows the ways that interactivity and transnationality are mined to amplify shifts in traditional understandings of family precisely because YouTube offers the fan base a medium of participatory culture.

The dependence on media that allow active participation through content creation applies to many quick media technologies. However, a similar dynamic is also at play in literary reading. In the case of epistolary fiction, a fictional genre which highlights and calls attention to the medium of transmission, a sense of kinship can emerge through the reader's recognition with the protagonists' identity negotiations. Silvia Schultermandl's chapter, "Literary Letters and IMs: American Epistolary Novels as Regulatory Fictions," argues that this is true for both quick and "slow" media (instant messages and letters). Her chapter examines the negotiation of national identity in American epistolary fiction by looking at the discursive shifts from letter writing in eighteenth-century novels to instant messaging in contemporary adolescent novels: both literary phenomena address young reading audiences and treat issues of femininity and female morality as conscripted by a US nationalist doctrine. While the novels mimic different media, the messages they impart are astonishingly similar, namely, that there is a consolidated idea what a "good" "American" "girl" does and how her communication with her friends can steer her in the right direction.

In both cases, whether mediated through quick media or prompted by the reader's recognition through the act of reading, interactivity facilitates means of renegotiating kinship ties across long and short distances and allows individuals to selectively identify with or disconnect from local, regional, global, but always inherently transnational communities.

Cyber-Alternatives to Lived Identities

Quick media technologies also offer the potential of negotiating online forms of kinship which are not easily attainable through embodied practices. They generally provide alternative means of identification that allow users to circumvent established systems of patriarchal and heteronormative control. In a study on the empowering potential of the Internet, Mehra, Merkel, and Bishop suggest that marginalized and minority groups including low-income families, members of the

LGBTQ communities, and African American women strategically use the Internet to access social privileges (education, knowledge, political agency, etc.) from which they are otherwise excluded. The availability of such alternative venues and their relative accessibility allows marginalized groups to gain a sense of empowerment in political, social, economic, and cultural contexts. That the Internet is "actually and potentially revolutionary for women and feminist activism" (Youngs 190), making these alternative forms of identity possible in the first place, is also central to the chapters in this section. In the context of queer Web-based communities, quick media technologies enable strategies of coming out which individuals may be hesitant to embrace in embodied practices. The transnational reach of chat rooms, for instance, also provides access to global queer communities primarily characterized by a shared sense of not national or cultural but sexual identity. In "Digital Diasporic Experiences in Digital Queer Spaces," Ahmet Atay explores queer social networking sites as "patched-together" and "digitally infused" alliances. Atay specifically takes an autoethnographic approach to unpacking the ways that virtual spaces provide opportunities that are radical revisionings of human interactivity across time and space and contextualizes them with more or less successful and fulfilling attempts at embodied practices of community building.

A similar case in point is the negotiation of ethnicity through quick media technologies, which enable Korean adoptees who grew up in white homes access to a form of kinship they did not develop with their adoptive families. Searches for birth parents and cultural origins yet unknown to them constitute alternative modes of identification and close an important cultural gap between the adoptees' socialization into white America and their Korean ancestry. In this vein, Kimberly McKee's "Claiming Ourselves as 'Korean': Accounting for Adoptees within the Korean Diaspora in the United States" examines the experiences of Korean children adopted by non-Korean American parents in navigating Korean American subjectivities through a message board devoted to Korean American identities. Her findings show that this online forum serves as a rich case study to explore the hybridized life of the transracial, transnational adoptee.

The resulting effects of maintaining both embodied and online kinship ties are hybridized and bifurcated identities that use quick media technologies to trace the lines of demarcation and potential contradictions between their own subjectivities and those of the communities with which they enter into contact.

Disembodied Connections

The use of quick media technologies not only enables alternative modes of identity negotiations but also substitutes practices of embodied and enacted identity. This is, of course, facilitated by the increasing possibilities of transgressing spatial and temporal boundaries in ways that make phenomena of multiple connectivity possible in the first place. Anthony Giddens, for instance, refers to such practices by introducing the term "disembedding," meaning a means to "live out" social activities and to extend them into farther-reaching realms, allowing users of quick media to pursue social relations over ever greater distances and providing access to otherwise inaccessible kinship ties. For Giddens, this "time-space distanciation" (90) epitomizes the ways in which modernity "increasingly tears space away from place by fostering relations between 'absent' others, locally distant from any given situation of face-to-face interaction" (18).

The productive and at the same time limiting effect of such increasingly disembedded encounters shapes kinship ties in unprecedented ways. Compared to older forms of communication before the onset of quick media, transnational encounters were more tenuous and rare, but perhaps also more significant: because the availability of such exchanges was limited, they were reserved for truly meaningful content; and through this scarcity, what was transmitted was automatically considered important. By looking at the shifts in communication practices within several generations of one family, one may trace considerable changes to the value and meaning of individual messages. For instance, May Friedman's chapter, "Shifting Terrain: Exploring the History of Communication through the Communication of My History," explores the evolution of family across immigration and technology. Investigating through a queer/transnational autoethnographic lens, it examines the limited access to communication formed by her parents' halting English love letters, which took weeks to cross the ocean, and reflects on the ways that present interfamilial communication has simultaneously become deeply dependent on emergent technology (both locally and transnationally) and hollowed by the same innovations.

At the same time, but in different contexts, quick media technologies allow for strategies of self-negotiation and kinship ties to which individuals may not otherwise have access. Being present while being away brings into sharper focus the dynamics of transnational identity, whereby individuals access different modes of self in different contexts. These possibilities are also explored in Isabella Ng's reflections on her family of origin and the maintenance of familial ties across transnational boundaries. In "Love Knows No Bounds: (Re)Defining Ambivalent Physical Boundary and Kinship in the World of ICTs," Ng considers both the strengths and limitations of Internet Communication Technologies on her families' capacity for closeness. Mixing autoethnography with communication theory, she comes to the conclusion that closeness and distance equally shape the sense of kinship in her family and notes eloquently that "We are together but apart."

The question of what gets lost in this increasing flow of data and to what degree abundant connectivity dilutes the quality of kinship ties applies equally to transnational individuals who are constantly travelling and those who seek to connect with family abroad. The impression that the closeness implied through quick media technologies may also become a form of constraint on personal freedom is an interesting side effect of this collapse of the time-space distanciation. Paradoxically, the seemingly constant availability in some cases also prompts feelings of disconnect and alienation. Travelling through remote areas with insufficient quick media infrastructure for extensive periods of time may highlight how a person is often thought to exist primarily through his or her online presence. This form of presence can be controlled, monitored, and traced to a degree where personal freedom becomes largely restricted. Foregrounding the artificial nature of online connectivity, as reflected in his chapter title, "The Internet Is Not a River: Space, Movement, and Relationality in a Wired World," Samuel Veissière asserts that "social media [has] provided a hungry void to channel people's fundamental urges to connect with one another in an anomic world but can never provide the real bodily platform to fulfil those desires." Bending traditional notions of analysis and certainty allows Veissière to reflect on these themes through his experiences as a mobile transnational subject navigating teaching, learning, parenting, and travelling through technology.

The 10 chapters of this book, together with the introduction, convey a variety of perspectives on the relationships between new media and transnationalism, viewed through analyses of various types of kinship. Importantly, the chapters do not all come to the same conclusions – some are more persuaded by the revolutionary potential of quick media than others, and each takes up themes of transnationality and kinship in different ways. Collectively, however, the chapters convey the timeliness and potency of new communication technologies as methods of identity construction in and beyond families and in and beyond borders.

WORKS CITED

Anderson, Benedict. *Imagined Communities: Reflections of the Origin and Spread of Nationalism.* 1983. London: Verso, 2007. Print.

Bauman, Zygmunt. *Modernity and Ambivalence.* London: Polity Press, 1991. Print.

Bornstein, Kate. *A Queer and Pleasant Danger.* Boston: Beacon Press, 2012. Print.

Briggs, Laura, Gladys McCormick, and J.T. Way. "Transnationalism: A Category of Analysis." *American Quarterly* 60.3 (September 2008): 625–48. Print.

Brinkerhoff, Jennifer M. *Digital Diasporas: Identity and Transnational Engagement.* New York: Cambridge University Press, 2009. Print.

Butler, Judith. *Gender Trouble: Feminism and the Subversion of Identity.* New York: Routledge, 1990. Print.

Butler, Judith, and Gayatri Chakravorty Spivak. *Who Sings the Nation-State? Language, Politics, Belonging.* London: Seagull, 2007. Print.

Doyle, Laura. "Toward a Philosophy of Transnationalism." *Journal of Transnational American Studies* 1.1 (2009): n.p. Web. 10 Feb. 2015.

Finney Boylan, Jennifer. *Stuck in the Middle with You: A Memoir of Parenting in Three Genders.* New York: Random House, 2013. Print.

Fischer, Hervé. *Digital Shock: Confronting the New Reality.* Montreal: McGill-Queen's University Press, 2006. Print.

Friedman, May. *Mommyblogs and the Changing Face of Motherhood.* Toronto: University of Toronto Press, 2013. Print.

Friedman, May, and Silvia Schultermandl. "Introduction." *Growing Up Transnational: Identity and Kinship in a Global Era.* Ed. May Friedman and Silvia Schultermandl. Toronto: University of Toronto Press, 2011. 3–18. Print.

Giddens, Anthony. *The Consequences of Modernity.* Stanford: Stanford University Press, 1990. Print.

Gurak, Laura, Smiljana Antonijevic, Laurie Johnson, Clancy Ratliff, and Jessica Reyman. "Introduction: Weblogs, Rhetoric, Community and Culture." *Into the Blogosphere: Rhetoric, Community and Culture of Weblogs.* Ed. Laura Gurak et al. Minneapolis: University of Minnesota, 2004. Retrieved from http://conservancy.umn.edu/handle/11299/172840. Web. 16 Aug. 2015.

Halberstam, Judith. "Automating Gender: Postmodern Feminism in the Age of the Intelligent Machine." *Feminist Studies* 17.3 (1991): 439–60. Print.

Hall, Stuart. "Cultural Identity and Diaspora." *Identity: Community, Culture, and Difference.* Ed. Jonathan Rutherford. London: Lawrence and Wishart, 1998. 222–37. Print.

Hays, Sharon. *The Cultural Contradictions of Motherhood*. New Haven, CT: Yale University Press, 1998. Print.

Hegde, Radha S., ed. *Circuits of Visibility: Gender and Transnational Media Cultures*. New York: NYU Press, 2011. Print.

Jenkins, Henry. *Confronting the Challenges of Participatory Culture: Media Education for the 21st Century*. Cambridge, MA: Massachusetts Institute of Technology Press, 2009. Print.

Kaplan, Caren, Norma Alarcón, and Minoo Moallem, eds. *Between Woman and Nation: Nationalisms, Transnational Feminisms, and the State*. Durham, NC: Duke University Press, 1999. Print.

Karlsson, Lena. "Desperately Seeking Sameness: The Processes and Pleasures of Identification in Women's Diary Blog Reading." *Feminist Media Studies* 7.2 (2007): 137–53. Print.

Khagram, Sanjeev, and Peggy Levitt. "Constructing Transnational Studies." *The Transnational Studies Reader: Intersections and Innovations*. Ed. Sanjeev Khagram and Peggy Levitt. New York: Routledge, 2007. 1–18. Print.

Kim, Hosu. "S/kin of Virtual Mothers: Loss and Mourning on a Korean Birthmothers' Website." *Mediating Moms: Mothers in Popular Culture*. Ed. Elizabeth Podnieks. Montreal: McGill-Queens University Press, 2012. 284–302. Print.

Kinser, Amber. "Mothering as Relational Consciousness." *Feminist Mothering*. Ed. Andrea O'Reilly. Albany: SUNY Press, 2008. 123–42. Print.

Ladin, Joy. *Through the Door of Life: A Jewish Journey between Genders*. Madison: University of Wisconsin Press, 2012. Print.

Laguerre, Michel S. "Digital Diaspora: Definition and Models." *Diasporas in the New Media Age: Identity, Politics and Community*. Ed. Adoni Alonso and Pedro J. Oiarzabal. Las Vegas: University of Nevada Press, 2010. 49–64. Print.

Lionnet, Francoise, and Shu-mei Shih, eds. *Minor Transnationalism*. Durham, NC: Duke University Press, 2005. Print.

Lorber, Judith. *Gender Inequality: Feminist Theories and Politics*. Oxford: Oxford University Press, 2010. Print.

Madianou, Mirca, and Daniel Miller. *Migration and New Media: Transnational Families and Polymedia*. London: Routledge, 2011. Print.

McLuhan, Marshall. *The Gutenberg Galaxy: The Making of Typographic Man*. Toronto: University of Toronto Press, 1962. Print.

Mehra, Bharat, Cecelia Merkel, and Ann Peterson Bishop. "The Internet for Empowerment of Minority and Marginalized Users." *New Media and Society* 6 (2004): 781–802. Web. 30 June 2015.

Nakamura, Lisa. *Cybertypes: Race, Ethnicity, and Identity on the Internet*. New York: Routledge, 2002. Print.

O'Reilly, Andrea. *Feminist Mothering*. Albany: SUNY Press, 2008. Print.
– *Mother Outlaws: Theories and Practices of Empowered Mothering*. Toronto: Women's Press, 2004. Print.
Park, Shelley. "Cyborg Mothering." *Mothers Who Deliver: Feminist Interventions in Public and Interpersonal Discourse*. Ed. Jocelyn Fenton Stitt and Pegeen Reichert Powell. Albany: SUNY Press, 2010. 56–76. Print.
Rich, Adrienne. *Of Woman Born: Motherhood as Experience and Institution*. 1976. New York: Norton and Company, 1995. Print.
Serfaty, Viviane. *The Mirror and the Veil: An Overview of American Online Diaries and Blogs*. Amsterdam: Rodopi Press, 2004. Print.
Shohat, Ella. *Taboo Memories: Diasporic Voices*. Durham, NC: Duke University Press, 2006. Print.
Shohat, Ella, and Robert Stam, eds. *Multiculturalism, Postcoloniality and Transnational Media*. New Brunswick, NJ: Rutgers University Press, 2003. Print.
Spivak, Gayatri Chakravorty."Cultural Talks in the Hot Peace: Revisiting the 'Global Village.'" *Cosmopolitics: Thinking and Feeling Beyond the Nation*. Ed. Pheng Cheah and Bruce Robbins. Minneapolis: University of Minnesota Press, 1998. 329–48. Print.
Stone, Linda. *Gender and Kinship: An Introduction*. 2nd ed. Boulder, CO: Westview Press, 2000. Print.
Wilson, Matthew W. "Cyborg Geographies: Towards Hybrid Epistemologies." *Gender, Place, and Culture* 16.5 (October 2009): 499–516. Web. 15 July 2015.
Youngs, Gillian. "Cyberspace: The New Feminist Frontier?" *Women and Media: International Perspectives*. Ed. Karen Ross and Carolyn M. Byerly. Malden, MA: Blackwell Publishing, 2004. 185–209. Print.

Between the Individual and
an Imagined Community

1 "I Talk to My Family in Mexico but I Don't *Know* Them": Undocumented Young Adults Negotiate Belonging in the United States through Conversations with Mexico

LAURA E. ENRIQUEZ

> I could go to Mexico but I couldn't come back. So I've been here for almost 25 years. I talk to my family in Mexico but I don't *know* them.
>
> Nancy Ortega

As undocumented immigrants, Nancy and an estimated 11.7 million other individuals living in the United States cannot visit their country of origin (Passel, Cohn, and Gonzalez-Barrera). The militarization of the US-Mexico border has increased the financial and psychosocial costs of undocumented entry and led to the permanent settlement of many undocumented immigrant families (Massey, Durand, and Malone). Having lived in the United States since she was six, Nancy, now 28 years old, explains that her young age of migration and inability to visit Mexico has limited her connection to the family members who remain there. Although she and other undocumented young adults cannot physically cross borders, they are present in and influenced by a "transnational social field" where social exchanges are made across borders, in part through media (Basch, Glick-Schiller, and Szanton-Blanc).

Globalization and technological advances have facilitated the growth of transnational contact among family members, contributed to the flow of economic and social remittances, and encouraged the development of transnational identities (Hernández-León; Levitt and Lamba-Nieves; Levitt and Waters; Massey, Goldring, and Durand; Morley; Panagakos and Horst; Smith; Vertovec). However, two critical gaps remain. First, many scholars tend to take the ability to cross borders for granted and focus on face-to-face contact. In fact, studies on the 1.5 and second generations tend to examine how visiting one's country of origin affects transnational ties and identities (Fouron and Glick-Schiller; Kasinitz et al.; Kibria; Louie; Rumbaut; Smith). Second, most research contends that individuals on both sides of borders are using increased opportunities for contact to build transnational lives and identities, although to varying extents. However, visits to or communication with individuals in the country of origin can make second-generation individuals feel that they do not belong by reinforcing feelings of marginalization and difference (Espiritu and Tran; Kibria). These gaps suggest that research on transnational identities needs to (1) account for the experiences of individuals who cannot easily cross borders and (2) further investigate how transnational contact may lead to disconnection. To this end, I explore the transnational connections that undocumented Mexican-origin young adults, like Nancy, develop with their family members and friends in Mexico.

Although quick media operate in borderless space to close the distance between individuals in different countries, we are living in an increasingly bordered world where transnational movement is limited by restrictive immigration laws. As a result, different types of transnational contact have emerged, as in the case of undocumented immigrants who must rely almost exclusively on media-facilitated, rather than face-to-face, contact to develop and maintain connections to their kin and imagined communities. This leads me to ask how 1.5-generation, undocumented young adults, who immigrated as children, build transnational ties when they cannot cross borders. How does media mediate relationship building? How might these new kinship experiences compare with those of first-generation undocumented immigrants who immigrated as adults? What implications does this type of contact have for their lives in the United States and the identities they develop?

Almost a fifth of undocumented immigrants in the United States are youth who arrived before the age of 16 and are currently under the age of 35 (Batalova and McHugh). Unlike adult immigrants, they have limited social and emotional ties to their country of origin because they migrated at a young age, have few memories of it, and are unable to visit due to their undocumented status. At the same time, they occupy a liminal legal status in the United States, where they have access to some rights but not others (Abrego, "Legitimacy, Social Identity"; Motomura). Although undocumented immigrant youth have the right to attend primary and secondary school, they begin to recognize the limitations associated with their immigration status as they transition into young adulthood and experience limited access to higher education, restricted employment opportunities, and curtailed physical mobility for fear of deportation (Abrego, "Incorporation Patterns"; Chavez; De Genova; Enriquez, "Educational Success"; Gonzales, "Learning to Be Illegal"; Gonzales, "On the Wrong Side"; Huber and Malagon; Lopez; Menjívar and Abrego). Seeking to combat these limitations, undocumented young adults draw on a lifetime spent in the United States to develop legal consciousness and claims-making strategies that allow them to assert their belonging and politically mobilize for positive immigration policies (Abrego, "Legal Consciousness"; Enriquez, "Building a Cross-Status Coalition"; Perry).

In light of the highly marginalized social world they live in, I find that the mediated nature of quick media affects how undocumented young adults are able to build relationships and express identification with their imagined families and communities. They must primarily draw on quick media to build and maintain transnational relationships with family members in their country of origin. Although they expect to experience a sense of kinship with these people because of their biological relationship, they find that media-facilitated contact is often not enough to construct this imagined family. Instead, their contact often reveals differences, rather than similarities, and disrupts their feelings of kinship. Yet quick media technology does offer a unique tool through which undocumented young adults can develop and express their identification with the broader imagined community of the United States.

Although previous research suggests that transnational contact enhances the development of a transnational identity, I demonstrate here that contact can also heighten a sense of disconnection and contribute to the development of a nation-based imagined community and identity. Specifically, quick media cannot bridge the gap created by a lack of pre-existing relationships and face-to-face contact with family members. As such, I argue that media-facilitated transnational ties are rich in possibility for claiming belonging in and asserting identification with a nation-based imagined community. In this case, undocumented young adults draw on conversations via quick media to imagine themselves as part of the host country (the United States), not necessarily the country of origin (Mexico) or a transnational diaspora. I first share a case study of one family's experience to demonstrate how media-facilitated contact has different effects on the transnational ties of first-generation immigrant adults and 1.5-generation immigrant children. I then discuss how, despite contact, the 1.5 generation struggles to build significant relationships with extended family members because of their young age of migration and limited pre-existing relationships. Finally, I demonstrate how this contact relays information to undocumented young adults that can be used to assert that they do not belong in Mexico but do belong in the United States. This suggests that, in a highly marginalized context, contact exclusively via media with family in the country of origin can actually weaken transnational or country-of-origin identification.

Methods

I draw on in-depth interviews with 90 undocumented and 29 recently legalized Mexican-origin young adults. Currently ages 20–35, these young adults are all 1.5-generation immigrants, having entered the United States before the age of 16. Most were significantly younger, with 40 per cent leaving Mexico before the age of 6 and another 40 per cent before the age of 10. Migrating during early childhood, the vast majority had not developed strong memories of Mexico, nor have they visited their extended families or country of origin for over 15 years. The sample includes relatively equal numbers of men and women and respondents who represent a variety of education levels, from not completing high school to having a postgraduate degree.

All respondents live in, and most grew up in, Southern California, home to one of the largest undocumented populations in the United States (Fortuny, Capps, and Passel). Drawing on the networks I developed during my previous fieldwork, I recruited respondents using a snowball sampling technique, where individuals received incentives for both participating in an interview and referring additional respondents. My 12 initial respondents and subsequent participants increased this sampling by referring family members, neighbours, former classmates, co-workers, and friends. All respondents have been assigned pseudonyms to protect confidentiality.

I conducted semi-structured interviews between November 2011 and August 2012. I asked respondents a range of questions addressing their migration histories, past and present experiences in the United States, and their overall feelings about citizenship and belonging. For this chapter, I draw mostly from their discussion of (1) their feelings of belonging and treatment in the United States, (2) their connections to and feelings about Mexico, and (3) their feelings about being deported or choosing to return to Mexico. Although I did not ask explicit questions about the type or amount of contact they have with family in Mexico, these issues emerged as respondents shared this information to illustrate their points. I also draw from six years of informal participant observation, conducted between 2007 and 2014, with undocumented young adults in Los Angeles, California.

Contact and Connection: Differences Between the First and 1.5 Generations

The Ortega family migrated to Los Angeles, California, from Puebla, Mexico, in 1989. Sara Ortega was 24 and looking for a better life for her two young children – Nancy, age 6, and Mauricio, age 5. Crossing the border without documents, they began a new life in the United States – Sara going to work in a garment factory and Nancy and Mauricio enrolling in school. As the years passed, they settled into their new life while trying to maintain contact with family in Mexico. Being undocumented, they could not travel back to visit, so Sara sought to maintain contact with her family through phone conversations. Though she often passed the phone to Nancy and Mauricio as they were growing up, they did not have many memories of their aunts, uncles, and grandparents. Their conversations were short and superficial, leaving little room to build significant relationships. As a result, Sara and her children experience these conversations differently. Specifically, her older age of migration allowed Sara to develop strong bonds with family members prior to migration, while the young ages of Nancy and Mauricio meant that they have few memories of their country of origin and the family members who remain there.

First- and 1.5-generation undocumented immigrants have different amounts and types of contact with their family members. Sara explains that she primarily contacts her parents and older sisters by phone once a week for one to two hours, wherein they discuss their lives, current events, and her aging parents' health. In addition, her mom sends her *paquetes* (courier-delivered packages) about once a month with candy, fresh local food (e.g., ears of corn, cheese, nuts), beauty products, pictures, and videos. She believes these provide "a way to not lose communication ... To find things to talk about." Alternatively, for most of their lives, Nancy and Mauricio's contact has consisted primarily of short, infrequent phone conversations with their grandmother. It was not until the past couple of years, when their aunts and uncles joined Facebook, that there were opportunities for additional contact. Nancy explains that she uses the chat and messenger functions to talk with her aunts and uncles "on Facebook maybe three or four times a week," and Mauricio says that "although we don't have any huge conversations there are definitely things that we post on our walls or share with each other … Old pictures and memories. One uncle who is a teacher shared a list of [Spanish-language] books I could use to teach ESL [English as a Second Language] students in my classes." Given the fact that younger generations are more familiar with technology and use quick media on a daily basis (Benítez), Nancy and Mauricio have adopted these media to somewhat increase their communication with family members in their country of origin.

Although technology has facilitated contact between the United States and Mexico, first- and 1.5-generation undocumented immigrants attach differing significance to this contact as a result of their age of migration. I was with the family one day when Sara received a *paquete* containing a DVD of a family wedding from a few weeks before. Watching the sea of faces on the video, she grew increasingly excited, shedding tears of joy and sadness as she pointed out family members to her children and recalled memories of them. Whereas her eyes were glued to the screen, Nancy, Mauricio, and their US-born sister Allie grew increasingly disinterested in the video, saying that they did not remember these people who they had not seen for over 20 years. Although Benítez points to the importance of visual contact through teleconferencing, Nancy and Mauricio's reactions suggest that even visual contact – through photographs, recorded videos, or live Skype conversations – may still not be enough to build a relationship when there are no memories to build upon. This indicates the importance of having pre-existing relationships in order to make this contact significant. Furthermore, Sara explains that despite her significant level of communication with family, she does not talk with her younger brothers because "I don't know them more than their voice … Because when I came here, they were small. So now I don't know how they are." Nancy builds on this idea, saying, "We're family and we care about each other but the connection is not the same. I know they are my family and that's the reason I talk to them. But there is not that bond. I make sure they are ok but there is not that connection." Both Sara and Nancy suggest that pre-existing memories and relationships are responsible for creating a bond upon which media-facilitated contact can then be used to maintain a relationship.

Thus, first-generation immigrants who migrated as adults are able to use this type of transnational contact to maintain connections, while 1.5-generation immigrants who migrated as children do not have the pre-existing memories needed to build significant relationships.

Contact also provides the Ortega family with information about life in Mexico. They all explain that their conversations tend to include information about current events, political elections, poverty in everyday life, and natural disasters. While this news does not strike Sara as out of the ordinary, Nancy and Mauricio tend to compare these realities to life in the United States. Mauricio says, "I don't feel any connection towards Mexico. I feel mostly the negative aspects of it – corruption, crime, conservatism – and I don't want to associate myself with it. I see this in the news but also in Facebook interactions. My uncle tagged me in a post about how this politician just bought votes and won an election." These conversations confirm media messages about Mexico and reinforce Mauricio's determination not to return. Thus, while contact may increase awareness about life in Mexico, it does not necessarily create a connection to the country of origin and can in fact create disconnection.

Contact also has different implications for how they imagine the possibility of returning to Mexico, either by choice or through deportation. Sara says, "If I return to Mexico, I will be fine with my older siblings" because they have maintained their relationships over the past 25 years. However, Nancy feels the opposite: "If I had to go back I would have to reconnect to my family ... Even though we do talk a lot, it's not the same as being around that person, growing up with that person." Mauricio adds, "I would feel like the people I would be living with were strangers ... All this Facebook stuff is better than nothing ... [But] because I don't plan on going back, those deeper conversations are not initiated and are not taking place." More recently, Mauricio began planning a visit to Mexico as part of his legalization process and has been in contact with family via email, Facebook, and telephone to plan the trip. Although he was now intentionally trying to build relationships, these conversations often ended in frustration and highlighted differences between himself and his family members, and between the United States and Mexico. This suggests that, despite one's intentionality, it is difficult to cultivate close connections primarily through media-facilitated contact.

The experiences of the Ortega family suggest that 1.5-generation, undocumented young adults have more limited interactions with family members in Mexico when compared to their first-generation parents. Nancy and Mauricio's experiences are fairly reflective of most respondents, who also discussed the significance of pre-existing relationships and messages about Mexico. Building on this case, the rest of this chapter explores the significance of young age of migration and how this limited contact transmits messages that affect 1.5-generation, undocumented young adults' feelings of belonging and identity development.

"I Don't Know Them": The (Im)Practicality of Building Relationships through Media

Most respondents have extended family members in Mexico, usually grandparents, aunts, and uncles. Their young age of departure means that most have limited memories of their country of origin and little recollection of pre-existing relationships with these individuals. Despite acknowledging that they are in contact with those who remain in Mexico, most respondents assert that this contact is limited and their relationships with kin are practically insignificant. This suggests that media-facilitated contact does not necessarily lead to deep relationships or transnational ties.

When speaking about family members who remain in Mexico, almost all respondents note that they have limited and tentative relationships. Juana Covarrubias migrated as a toddler and has no memories of the family members she has not seen for over 20 years. She explains, "I don't know them. I've never met them. So it would be hard to even feel comfortable." Despite the fact that she knew her family as a young child, Juana does not have any personal memories of them. In addition, infrequent phone conversations and updates about family in Mexico do not provide enough information for her to feel like she has knowledge of their lives or a significant relationship with them. Ivan Cardenas, who left Mexico at the age of four, espouses similar feelings, even though he consistently communicates with family in Mexico. He says,

> They haven't seen me in over 20 years. We talk on the phone like probably once a month. My grandma's always talking about how she misses me, how she remembers me when I was there. It's sad though because I don't remember her ... If my grandparents were to be walking [by here] right now, I wouldn't recognize them. I wouldn't know them. Everything over there is just a memory.

Although his age of migration allowed him to develop a few memories of Mexico, he notes specifically that these do not enable him to recognize his grandparents in the present. These memories are not enough to build a relationship on since their conversations tend to be grounded in the past. Further, the few ties that they have, especially to grandparents who took care of them prior to migration, can disappear as older generations pass away. Daniela Sanchez, who migrated at the age of four, explains that she has tried to maintain relationships with her family over the phone and more recently via Skype, but "people have passed away. Like my dad's mom passed away last year and my mom's dad passed away this Christmas. I have like maybe thirty seconds of where I can imagine them and remember them and that's it." Despite her attempts to build relationships, the natural passing of time cut off the few relationships she had been able to maintain with the family members she remembered the most. Together, Juana's, Ivan's, and Daniela's experiences suggest that, even considering the amount of familial contact they have now, they do not feel like they have been able to build and maintain significant relationships. This can largely be attributed to the fact that they did not have much previous significant contact and memories to build upon.

Even with limited memories, many respondents maintained some contact, usually through the phone calls their parents made, and some even aspired to develop relationships with their family. Zen Cruz, who was brought to the United States at the age of seven, has vivid memories of family members he left behind and now actively attempts to keep in contact with some of them through Facebook. He admits though,

> I have a lot of family in Mexico. And again, I don't know them. I mean, I know them, I know who they are, but I don't really *know* them because they live so far away and I can't see them. And we'll talk every now and then through Facebook, but it's not the same. I mean, they have their own life, and they have their own identities out there, and it's difficult to really connect with them.

Zen was one of the few respondents who actively pursued relationships with family members in Mexico. Despite his attempts to build these ties, he suggests that it is difficult to connect with family in Mexico because their lives are so different. While these Facebook conversations provided him with information about his family, they also showed him that his identity and his lifestyle are different from those of his family members. In addition, the experiences of recently legalized respondents suggest that the significant lapse of time between their departure and return makes it hard to reconnect. After 18 years, Leslie Fuentes received her legal permanent residency and was able to travel to Mexico and meet her family. She recalls,

> I had talked briefly on the phone [to my family]. I had imagined them in my head my entire life ... I longed to be able to know them and to get a sense for what their lifestyle was like, what my life might've been like if my parents hadn't come over here ... A lot of my cousins, we still talk all the time. On Facebook we chat and I still would love to get to know them better [but] their lives are so crazy and I'm all busy over here.

Leslie has both a desire to build connections and an ability to build on recent face-to-face contact, but she struggles to use quick media to build these relationships because it only offers a small stream of contact that is difficult to deploy amidst a busy schedule. In the absence of previous relationships and in a busy world, media-facilitated contact offers an opportunity for relationship building but the limited contact it offers can be insufficient for building significant relationships.

Although most respondents have some level of contact with family back in Mexico, they struggle to build relationships out of their fading memories. This is especially true when attempted through media-facilitated contact that only allows for the transmission of a limited stream of information. This suggests that, on its own, transnational contact may not be enough to produce transnational identities, especially when this contact is being used to build, rather than maintain, a relationship.

"It's Pretty Bad Over There": Claiming Belonging in the United States

Limited memories and relationships with family members can also be extended to suggest a lack of relationship to the country of origin. Cameron Peña discusses why he does not feel like he would belong in Mexico if he returned or was deported there:

> It seems challenging. My grandma's down there and I have a lot of family down there. But ... I left at an early age [four years old] so I don't know them. I remember my grandmother and I know who she is, and certain aunts and uncles, but I don't really have an attachment to Mexico.

For Cameron, and for most respondents, a lack of connection to people translates into a lack of connection to the country. This suggests that many undocumented young adults have difficulty forming a transnational identity because their early age of migration and limited personal memories restricts their pre-existing transnational ties. Yet respondents with significant memories and relationships also sought to deny their significance. For example, Irene Correas, who came to the United States at the age of six, suggests that she has no connections in Mexico in spite of the fact that her brother now lives there after being deported. She explains her lack of connection to Mexico: "I can't really say it's my home. It might be my native country, but I can't really live there. I don't really have friends over there … I wouldn't know where to go." Though she speaks to her brother frequently and had a strong relationship with him prior to his departure, Irene insists that she has little contact with and knowledge of Mexico. Even with their differing levels of contact with family in Mexico, Cameron and Irene both distance themselves from their country of origin by citing a lack of connection and belonging. This suggests that contact may do little to promote a connection to the country of origin, which drastically inhibits their ability to develop transnational identities or imagine themselves as part of a transnational or country-of-origin community.

Media-facilitated contact may not help 1.5-generation, undocumented young adults build relationships with family in Mexico, but it does serve as a source of information for identity formation. However, rather than encouraging and fostering transnational identities as the literature suggests, these conversations tell undocumented young adults that they do not belong in Mexico and help them imagine themselves as members of the US community. Most respondents cite conversations with family and friends in Mexico about violence, limited opportunities, and their inadequate cultural knowledge to demonstrate that they would neither belong nor thrive in Mexico. Even though these messages are decontextualized and are likely blown out of proportion, these conversations highlight differences and reinforce US-based identities and lives, instead of forging transnational ones.

Violence and Danger

The mass media contributes to a problematic and one-dimensional image of Mexico as a dangerous and violent country (Cave; Fantz; Shoichet). Ignacio Nuñez, who migrated at age 11, explains how these media images, however debatable, discourage him from wanting to return to Mexico after more than 20 years in the United States: "I think it's pretty bad over there. I see a lot of difference, especially when you see the news. And they're talking about Mexico. They're doing this and this ... All the people they get scared to go back. It's too dangerous over there ... I belong over here." Drawing on mass media messages about Mexico as plagued by drug violence, kidnapping, rape, and murder, respondents like Ignacio assert that it is too dangerous to return to Mexico and then explicitly use this information to claim that they belong in the United States.

These media images and feelings are validated by contact with family and friends who provide additional information about the danger and violence in Mexico. Raul Robles, who migrated at age 16 with his older brother, asserts that it would be challenging to belong in Mexico if he returned. He recounts the conversations he has with his parents and friends whom he has not seen since he left six years ago.

> The Mexico I left wasn't like that ... When I talk to my parents over the phone, the things that they're saying to me ... I had a conversation with a friend of mine, over Facebook, we were chatting ... And, in a way, I'm glad I left ... He was telling me about what is happening in Acapulco, where I spent most of my life, where my family is from ... They put the central market on fire because of drug retaliation ... Knowing that people that went to school with me, during elementary, middle school and high school have been killed because of the drug violence.

Unlike most respondents, Raul has strong pre-existing relationships with family and friends in Mexico because of his later age of departure. Though he is using phone calls and Facebook to maintain these ties, these conversations reinforce how Mexico has changed and that he would be in danger if he returned. Similarly, Irene Correas's contact with her extended family and deported older brother has directly exposed her to the violent images of Mexico. She explains, "[My brother] was kidnapped recently ... That makes me not want to go back to Mexico anymore ... I guess people think he has money, so they kidnapped him on Friday night. They just released him last Monday, and we had to come up with like $6,000. It was really scary." These experiences suggest that contact with family and friends provides opportunities for directly reinforcing negative media images of danger and violence in Mexico.

Worried about their physical safety upon returning to Mexico, respondents use the information provided by transnational media conglomerates and contact with family and friends to assert belonging in the United States and claim a right to remain. For example, Gilbert Morales draws on his conversations with family and media images to assert that he does not belong in Mexico "because [of] all the news I've been hearing. I don't even want to go to Mexico, not even for vacation right now ... [because of] all the murders, violence and everything." He continues to draw on these images to make a connection to the United States, saying, "When I dream that I'm in Mexico, I'm crying ... What am I doing here? What am I gonna do? I don't want to go back to Mexico because my life is right here." Like most respondents, Gilbert desires to remain in the United States, the country where he grew up and has spent the majority of his life. Drawing on these violent images and asserting a fear for his return to Mexico allows him to further support his claim for the right to remain in the United States. On the other hand, Joanna Salas expresses a sense of connection to Mexico: "I talk to my grandparents every time my dad calls. My dad always shows us pictures. Movies they bring from over there. I see Spanish TV." Even while she asserts that these connections are significant to her identity as a Mexican, she rejects the idea of returning to Mexico, saying, "I do think I belong in my country [of Mexico] though I'm scared because of everything that's going on there like the violence and so on." In the few cases where respondents feel that they are able to build connections to their country of origin through this media-facilitated contact, the information they receive can still dissuade them from returning. This suggests that other messages, in this case those from mass media, can overpower transnational contact with family. As a result, transnational conversations and media images can produce feelings of disconnection with a country of origin and provide fodder for asserting belonging in the host country.

Limited Opportunities

Similar to the narrative of violence is an image of Mexico as a land of limited opportunities. Initial messages about opportunities are transmitted through migration histories as parents tell their children that they migrated to the United States for the opportunities it offered. Each family's rendition of the American dream simultaneously informs respondents that there is a lack of opportunities in Mexico, regardless of whether this is true. Marisol Salas, who was brought to the United States at the age of three with her two siblings, explains that her limited opportunities as an undocumented young adult are still better than what might be available for her in Mexico:

> I know my parents brought me here to get a better life. If I was to be in Mexico, I wouldn't be speaking English or be knowing other stuff. I see the life that they live and it's really sad. There are a lot of children that live over there that don't have shoes or clothes. They're really dirty and it's really sad. Sometimes they don't even have food. So sometimes I really appreciate my dad that he worked hard for us to come here.

With these highly skewed perceptions, she explains that she knows these things about Mexico "because we have seen videos. Sometime they [my parents] show things on TV or on the Internet." Despite the fact that she and her undocumented siblings struggled to complete high school and secure stable employment, Marisol believes that her opportunities in the United States are better than in Mexico because of the poverty-stricken reality she sees on videos and messages sent by her family in Mexico and the superficial images she sees in the mass media. This imagery reinforces a desire to remain in the United States.

Though these parental narratives about opportunities and differences in lifestyles inspire appreciation for life in the United States, they also shape how undocumented young adults react to conversations with family members in Mexico about their return. Yazmin Flores explains how conversations with her grandmother strengthen her aversion to returning to Mexico and reinforce her desire to remain in the United States:

> I know that if I'm ever deported I'm not going to make it in Mexico … My grandma, she just tells me, "Come on, if you get deported, you already have your room." I'm like, "I don't wanna go." And then the bedrooms don't have – cause I've actually seen pictures – they don't have glass. It's just the window and the curtain. So I'm like, "Uh-uh! [No.] It's scary." … I know I'm not going to be that strong to be there even though I'm going to have a lot of family members.

Even though her grandmother seeks to assure her that her deportation would not be catastrophic, the information she receives about life in Mexico through their transnational contact – phone conversations and shared photos – assures Yazmin that she would not survive if she were to return to Mexico. As a result, she draws on this contact, and the information it provides about lifestyle differences, to assert that she should not be deported and should be allowed to legalize her status and remain in the United States. In this way, transnational contact further distances her from her country of origin and alerts her to the significant differences in opportunity.

Inadequate Cultural Knowledge

In addition to creating a fear of life in Mexico, conversations with family and friends convince respondents that they do not have the cultural knowledge necessary to survive in Mexico. Most commonly, respondents believe that their Spanish-language skills are inadequate for navigating life in Mexico should they return. This message is transmitted to many through conversations with their family in Mexico and reinforced when they tried to speak Spanish with Mexican-origin adults who migrated at a later age. Leo Campos explains,

> I'm American. If you're going to toss me back, you might as well put a bullet in my head because I will not survive … I had a conversation with one of my aunts in Mexico the other day and I couldn't think of half the words. I got off the phone and handed it back to my mom. My Tía [aunt] straight [out] said, "Poor boy! He doesn't speak Spanish well." Because I can't. I sound like an American trying to speak Spanish … I don't sound Mexican at all … When I said bye I said, "*Tenga buen día* [literally: Have a nice day]." I thought that's what you say, but that means something else.

While Leo's conversations with his aunt taught him that he was different than Mexicans living in Mexico, he also uses this knowledge to assert that he is American and to claim a right to remain in the United States. Similar to Leo, other respondents made claims to remain in the United States by citing that they "sound funny" when speaking Spanish, are "losing my Spanish," and do not "get it" when they talk to family in Mexico because they speak too fast or use slang. In the rare event that a respondent had returned to visit,[1] the disconnect was even

---

1 This was usually 15–20 years previous to the interview, when they were still children. At the time, the US-Mexican border was less militarized and undocumented border crossing was easier and less costly.

clearer. Tanya Diaz recalls her experiences in Mexico upon returning for her grandfather's funeral when she was 12: "It was horrible because I forgot my Spanish somewhat ... So I was called *gringa* [a derogatory word for foreigner]. I was like, where am I accepted? I don't belong here [in Mexico]. I don't belong over there ... [But] I know I belong here [in the United States]." Like Leo, Tanya asserts that she belongs in the United States by drawing on these conversations with her family from almost 15 years ago. Not only did they teach her that she doesn't know Spanish well enough, but they instilled in her a dread of returning to Mexico. For many respondents, conversations with family in Mexico highlight differences and lead them to feel that they will not be able to communicate with others and navigate Mexican society if they return.

As with language, transnational conversations transmit information about shifts in popular culture. Patricia Santamaria recounts the conversations she has with friends she has not seen for six years, since leaving Mexico at the age of 15: "I stay in touch with my friends and I talk to them all the time. They have different interests. They have done a lot of things that I haven't done. They know a lot of things that I don't know. So obviously I wouldn't be on the same page if I go back." These conversations force Patricia to realize that she is becoming increasingly different from her peers in Mexico; as the years pass, their lives go on and their interests and cultural references shift away from one another. Although her older age of migration ensures that she has some pre-existing knowledge of Mexico, most respondents who migrated at younger ages do not even have this previously cultivated cultural knowledge. For example, Eva Santiz explains, "I don't feel like I'm fully Mexican because I don't know there. I don't know anything. I don't know any traditions. I don't know where I used to live. I don't know. I'm blank. Sometimes I feel that my [US-born] boyfriend's more Mexican than me cause he knows more about it." Whereas Patricia's previous cultural awareness has changed over time, Eva suggests that her complete lack of cultural and institutional knowledge prevents her from fully identifying with Mexico. These conversations raise individuals' awareness that they lack the cultural knowledge needed to navigate their country of origin and contribute to their feelings of exclusion from Mexico.

Overall, conversations with family and friends in Mexico do not produce transnational identities but rather serve as a source of information that encourages undocumented young adults to develop and assert US-based identities and belonging. These conversations produce images of Mexico as a violent place with limited opportunities and highlight wide cultural gaps. As a result, contact can produce feelings of disconnection and exclusion, which affects how individuals conceptualize their imagined community. As such, media-facilitated transnational contact can actually produce a national rather than a transnational identity, especially in cases where structural exclusion is high.

Conclusion

Contact is often assumed to mean connection, and transnationalism has come to signal the limited significance of nation-state boundaries (Bloemraad; Massey, Goldring, and Durand; Smith). I argue that this is not necessarily true in the case of 1.5-generation, undocumented young adults in the United States, a highly marginalized population with limited legal rights and a restricted ability to physically move across nation-state borders. I find that undocumented young adults who immigrated to the United States as children have few memories to build upon, so media-facilitated transnational contact only produces superficial relationships. This suggests that media-facilitated contact must be coupled with physical contact, as in the experiences of first-generation undocumented immigrants, in order to develop a transnational imagined community.

Although quick media technology has opened up more avenues for contact in borderless virtual spaces, I contend that this type of contact is not enough to produce a transnational identity in a global society that prevents face-to-face contact with restrictive immigration laws that prevent geographic mobility. Just as Benítez demonstrates the emotional importance of visual connections through teleconferencing, I find that face-to-face contact is critical for initiating and sustaining relationships across borders. This suggests that the mediated nature of quick media, as opposed to face-to-face contact, can influence identity formation processes and provide fodder for identity claims by limiting, rather than promoting, transnational identities. The limited potential of quick media contact on its own suggests that all types of contact are not equal and future research on transnational identities needs to be more specific about the meaning of "continuous regular contact across national borders" (Portes, Guarnizo, and Landolt 217).

Despite the superficial nature of these transnational relationships, quick media do offer a unique tool through which undocumented young adults can develop and express their identification with an imagined community. In particular, these conversations relay messages about life in Mexico, which undocumented young adults use to claim belonging in and assert identification with the United States, rather than the country of origin. These conversations provide information about the violence, limited opportunities, and differing cultural norms in Mexico and reinforce negative popular media messages. Undocumented young adults then draw on this information as they construct their chosen identities and negotiate their belonging to their imagined community in the United States. This suggests that media-facilitated contact can produce a national, not a transnational, identity in some cases. Further, the emergence of US-based national identities in this particular case suggests that nation-based identities may be more common when individuals experience highly exclusionary and marginalizing circumstances. I contend that this occurs in the case of undocumented young adults because they are unable to cross nation-state boundaries and are subject to liminal legality in the United States, which forces them to develop innovative ways to assert their belonging in their imagined community.

WORKS CITED

Abrego, Leisy J. "'I Can't Go to College Because I Don't Have Papers': Incorporation Patterns of Latino Undocumented Youth." *Latino Studies* 4.3 (2006): 212–31. Print.

– "Legal Consciousness of Undocumented Latinos: Fear and Stigma as Barriers to Claims-Making for First- and 1.5-Generation Immigrants." *Law & Society Review* 45.2 (2011): 337–70. Print.

– "Legitimacy, Social Identity, and the Mobilization of Law: The Effects of Assembly Bill 540 on Undocumented Students in California." *Law & Social Inquiry* 33.3 (2008): 709–34. Print.

Basch, Linda, Nina Glick-Schiller, and Cristina Szanton-Blanc. *Nations Unbound: Transnational Projects, Postcolonial Predicaments, and Deterritorialized Nation-States.* Langhorne, PA: Gordon and Breach, 1994. Print.

Batalova, Jeanne, and Margie McHugh. *Dream vs. Reality: An Analysis of Potential Dream Act Beneficiaries.* Washington, DC: Migration Policy Institute, 2010. Web. 14 Dec. 2013.

Benítez, José Luis. "Transnational Dimensions of the Digital Divide among Salvadoran Immigrants in the Washington DC Metropolitan Area." *Global Networks* 6.2 (2006): 181–99. Print.

Bloemraad, Irene. "Who Claims Dual Citizenship? The Limits of Postnationalism, the Possibilities of Transnationalism, and the Persistence of Traditional Citizenship." *International Migration Review* 38.2 (2004): 389–426. Print.

Cave, Damien. "Wave of Violence Swallows More Women in Juarez, Mexico." *New York Times* 23 June 2012. Web. 14 Dec. 2013.

Chavez, Leo. *Shadowed Lives: Undocumented Immigrants in American Society.* Fort Worth, TX: Harcourt Brace, 1998. Print.

De Genova, Nicholas P. "Migrant 'Illegality' and Deportability in Everyday Life." *Annual Review of Anthropology* 31 (2002): 419–47. Print.

Enriquez, Laura E. "'Because We Feel the Pressure and We Also Feel the Support': Examining the Educational Success of Undocumented Immigrant Latina/o Students." *Harvard Educational Review* 81.3 (2011): 476–500. Print.

– "'Undocumented and Citizen Students Unite': Building a Cross-Status Coalition through Shared Ideology." *Social Problems* 61.2 (2014): 155–74. Print.

Espiritu, Yen Le, and Thom Tran. "'Viet Nam, Nuóc Toi' (Vietnam, My Country): Vietnamese Americans and Transnationalism." Levitt and Waters 367–98.

Fantz, Ashley. "The Mexico Drug War: Bodies for Billions." *CNN* 2012. Web. 14 Dec. 2013.

Fortuny, Karina, Randy Capps, and Jeffrey S. Passel. *The Characteristics of Unauthorized Immigrants in California, Los Angeles County, and the United States.* Washington, DC: The Urban Institute, 2007. Web. 14 Dec. 2013.

Fouron, Georges E., and Nina Glick-Schiller. "The Generation of Identity: Redefining the Second Generation within a Transnational Social Field." Levitt and Waters 168–208.

Gonzales, Roberto G. "Learning to Be Illegal: Undocumented Youth and Shifting Legal Contexts in the Transition to Adulthood." *American Sociological Review* 76.4 (2011): 602–19. Print.

– "On the Wrong Side of the Tracks: Understanding the Effects of School Structure and Social Capital in the Educational Pursuits of Undocumented Immigrant Students." *Peabody Journal of Education* 85.4 (2010): 469–85. Print.

Hernández-León, Rubén. *Metropolitan Migrants: The Migration of Urban Mexicans to the United States.* Berkeley: University of California Press, 2008. Print.

Huber, Lindsay Perez, and Maria C. Malagon. "Silenced Struggles: The Experiences of Latina and Latino Undocumented College Students in California." *Nevada Law Journal* 7 (2007): 841–61. Print.

Kasinitz, Phillip, et al. "Transnationalism and the Children of Immigrants in Contemporary New York." Levitt and Waters 96–122.

Kibria, Nazli. "Of Blood, Belonging, and Homeland Trips: Transnationalism and Identity among Second-Generation Chinese and Korean Americans." Levitt and Waters 295–311.

Levitt, Peggy, and Deepak Lamba-Nieves. "Social Remittances Revisited." *Journal of Ethnic and Migration Studies* 37.1 (2011): 1–22. Print.

Levitt, Peggy, and Mary C. Waters, eds. *The Changing Face of Home: The Transnational Lives of the Second Generation.* New York: Russell Sage Foundation, 2006. Print.

Lopez, Maria Pabón. "More Than a License to Drive: State Restrictions on the Use of Driver's Licenses by Noncitizens." *Southern Illinois University Law Journal* 29 (2004): 91–128. Print.

Louie, Andrea. "Creating Histories for the Present: Second-Generation (Re)Definitions of Chinese American Culture." Levitt and Waters 312–40.

Massey, Douglas S., Jorge Durand, and Nolan J. Malone. *Beyond Smoke and Mirrors: Mexican Immigration in an Era of Economic Integration.* New York: Russell Sage Foundation, 2002. Print.

Massey, Douglas S., Luin Goldring, and Jorge Durand. "Continuities in Transnational Migration: An Analysis of Nineteen Mexican Communities." *American Journal of Sociology* 99.6 (1994): 1492–533. Print.

Menjívar, Cecilia, and Leisy J. Abrego. "Legal Violence: Immigration Law and the Lives of Central American Immigrants." *American Journal of Sociology* 117.5 (2012): 1380–421. Print.

Morley, David. *Home Territories: Media, Mobility and Identity.* London: Routledge, 2000. Print.

Motomura, Hiroshi. "The Rights of Others: Legal Claims and Immigration Outside the Law." *Duke Law Journal* 59 (2010): 1723–86. Print.

Panagakos, Anastasia N., and Heather A. Horst. "Return to Cyberia: Technology and the Social Worlds of Transnational Migrants." *Global Networks* 6.2 (2006): 109–24. Print.

Passel, Jeffrey S., D'Vera Cohn, and Ana Gonzalez-Barrera. *Population Decline of Unauthorized Immigrants Stalls, May Have Reversed.* Washington, DC: Pew Hispanic Center, 2013. Web. 14 Dec. 2013.

Perry, Andre M. "Toward a Theoretical Framework for Membership: The Case of Undocumented Immigrants and Financial Aid for Postsecondary Education." *Review of Higher Education* 30.1 (2006): 21–40. Print.

Portes, Alejandro, Luis E. Guarnizo, and Patricia Landolt. "The Study of Transnationalism: Pitfalls and Promise of an Emergent Research Field." *Ethnic and Racial Studies* 22.2 (1999): 217–37. Print.

Rumbaut, Rubén G. "Severed or Sustained Attachments? Language, Identity, and Imagined Communities in the Post-Immigrant Generation." Levitt and Waters 43–95.

Shoichet, Catherine E. "Mexico: 13 Police Officers Arrested in Bust of Kidnapping Gang." *CNN* 9 October 2013. Web. 14 Dec. 2013.

Smith, Robert Courtney. *Mexican New York: Transnational Lives of New Immigrants*. Berkeley: University of California Press, 2006. Print.

Vertovec, Steven. "Transnationalism and Identity." *Journal of Ethnic and Migration Studies* 27.4 (2001): 573–82. Print.

2 "Learning and Practising Democracy": Digital Diasporas and Negotiating a Transnational Civil Society

M. TINA ZARPOUR

Introduction

The seeds of this research are intertwined with my growing up years – on the fringes of Iranian culture as a person with a trifurcated identity (Iranian, Mexican/South Texan, American). During many long evenings of *meyhmooni* (Iranian social visits with food) at the homes of other Iranians and hearing daily telephone conversations my Iranian father had with his friends, I came to understand that the topic of conversation was politics – Iranian politics – always heated and passionate. Fast-forward two or three decades, and the issues and people change but the overall scene remains the same.

As the product of mixed parentage – an Iranian father and Mexican American mother – my siblings and I were rarities, especially in Texas, where I did the majority of my growing up. My family's background was set apart for another reason. My parents met in San Antonio, Texas, when my mother (escaped from her tiny Texas border town, where some of the streets are still not paved) was working as a secretary at Lackland Air Force Base and my father, a pilot with the Iranian Imperial Air Force, was sent by the Iranian government for training in the late 1950s. That was in a different era of diplomatic relations between the United States and Iran – friends then, enemies now! My parents married in 1960 and moved to Iran. All three of us children were born there, unlike most of the other Iranian "halfies" I might encounter in the United States.

California was always like a dream to me, with its vibrant landscaping, blue skies, and the Pacific Ocean. It was my parents' favoured vacation spot because we could stay with extended family. In Texas, we were fairly isolated from other Iranians. In San Diego, you can be a tourist picnicking on the green grass enjoying the cliffs and wildlife at La Jolla Cove, and not walk 10 feet without hearing someone speak Farsi, the most common language spoken in Iran.

Contrary to the dominant and prevailing tradition of anthropological fieldwork, I did not travel away, then return from some other place, to conduct fieldwork. Though I did not grow up in San Diego, or California at all for that matter, I have had extended family there for years and it is now my home as an adult parent, raising children. I have lived in San Diego for five years. Because of this move, the repercussions of which I cannot always grasp, the edges of fieldwork are blurred during the moments of heightened interaction I have described.

My partial insider status to Iranian culture as a member of its 1.25-generation (Rumbaut 1167) population allows me to understand impassioned political debate as bigger than just my dad and his friends – Iranians *seem* to be more political. One reason may be that Iran's repressive politics and policies towards its citizens initially politicize them, wherever they find themselves. This was certainly the viewpoint of one of my informants. Another possibility is the collective trauma of homeland displacement as a result of the 1979 Iranian Revolution puts you in the same boat as your compatriots, and for the most part you likely share distaste for your homeland government, though your ideologies may differ.

My insider status within American culture allowed me to understand that politics was one thing you just did not talk about with your friends in the United States. Discussing politics and religion among a circle of people is taboo – it might invite arguments or create unnecessary tension and division among friends. Friends' or acquaintances' political leanings and ideology have sometimes been such a dark hole that I have often been surprised to later find out where they fall on the political or ideological spectrum.

If the first reason for developing a research question was based on personal experience, the second impetus is observational and evolved out of witnessing with outrage and sadness how the Iranian government cracked down on protesters in the streets and other dissidents in the aftermath of the June 2009 re-election of Iranian president Mahmoud Ahmadinejad. Yet I also observed how the political turmoil in Iran ignited the Iranian diaspora around a common cause. I saw previously apathetic Iranians participating in rallies and marches throughout Europe and the United States, and circulating anti-government messages through Facebook, YouTube, and other virtual means of communication. My Iranian relatives and friends were genuinely engaged and excited about post-election events. It was as if they saw a rip in the shroud with light peeking through. Among some exiles, I heard hope that an eventual return to Iran would be possible soon.

This chapter looks to "digital diasporas" (Laguerre) and their potential in fashioning a transnational civil society for dispersed migrants. It is derived from a larger study focused on examining Iranian immigrant political agency in San Diego, California, and looked to the fallout of the 2009 contested re-election of President Ahmadinejad as a critical moment in Iranian diasporic and transnational politics.

In the larger study I used the rhetorical device of "political talk," which encompasses politically and civically oriented discourse, action, and ideology. I followed political talk as it presented itself in two locations within the public sphere: through the life course of Iranian Americans and through discourse in online forums. Here I will be elaborating on the discourse encountered in online forums and the online activity reported by informants. According to Habermas's definition, the public sphere mediates between the private sphere and the sphere of public authority (the realm of the state, government, ruling class) (30). Nancy Fraser rearticulates the public sphere as "a theater in modern societies in which political participation is enacted through the medium of talk" (57) to move from the idea of one public sphere to numerous "counterspheres" and multiple and discrete places where "talk" occurs (Rai 32). The public sphere is an arena of social life where people come together and discuss problems. It is the place from which political action springs. The public sphere stands outside the private sphere of the home, where *meyhmooni* takes place. Among the public sphere, organizations and associations – virtual and material – are the vehicles through which political participation ensues. I used a combination of conventional ethnography (participant observation, informal interviews, life history interviews) and virtual ethnography to help develop a typology of political and civic action among Iranian migrants.

Constructing a Diasporic Public Sphere in Cyberspace

Soheyl Amini points out that Iranians inside and outside Iran engage in cyberspace in high numbers. Inside Iran, the Internet has become a powerful tool of the opposition, aiding burgeoning social movements.

Iranian dissident groups have employed online activism to broaden their reach and circumvent established propaganda mechanisms of the Islamic Republic of Iran. They can directly exchange information and mobilize with other social movements, as well as develop "solidarity and sympathy around the globe" that would have been impossible with traditional means of communication (Rahimi and Gheytanchi, n.pag.).

Within the diaspora, the Internet satisfies a number of needs for Iranians. It provides a forum for intergenerational dialogue and a place for the second generation to learn about Persian expressive culture. Thanks to the Internet, Iranians in the diaspora can purchase products not available in their host country. Iranian entrepreneurs around the world advertise their goods and services through online magazines. The Internet serves as a site to engage in political debate and argumentation on a transnational scale, and information and debate flow across the globe. It is also a space to share intimate details about oneself that Iranian culture's norms about privacy would not traditionally allow. For Iranian immigrants, cyberspace constitutes an alternative territory where transnational community can be constructed easily and inexpensively (Graham and Khosravi 227–8). Werbner's articulation of a "diasporic public sphere" (12) is useful to visualize the flows of information and ideas between different media and from media to different structures within the public sphere. Werbner conceives of a diasporic public sphere "in which different transnational imaginaries are interpreted and argued over, where aesthetic and moral fables of diaspora are formulated, and political mobilization generated, often in response to global media events" (12).

Diaspora and cyberspace are conceptually linked. First, both are forms of displacement. People in a diaspora have to construct a social context for themselves that transcends their physical location. Second, diaspora and cyberspace are linked through the notion of "community" (Bernal 661). New forms of social belonging have arisen from both advances in communication technology and the geographic mobility of populations. Bernal, in her analysis of Eritreans in diaspora and cyberspace, finds that violence and conflict emerge as a central dynamic. Conflict is both destructive and productive of community and identity in the public sphere (662). Contrasting her analyses with Benedict Anderson's assertion that the imagined nation was intertwined with the medium of the newspaper, Bernal suggests that new media and new conditions of transnational migration and globalization are altering the lived experience of citizenship, community, and nationalism, as well as the ways in which these can be collectively imagined (673).

In a similar vein, Rai found through discourse encountered in electronic bulletin boards that the Indian diaspora is "being written and re-written in the interstitial space" (53). This represents a veritable "countersphere" that pluralizes subject positions in its heterogeneity. Thompson sees that new media used by ethnic minorities challenges the assimilation model and changes the paradigm of an identity with a single nation-state to a fragmented, hybridized spectrum of identities (417). Online discourse among diaspora members is an "ever-emerging text" where "no one is in control" and there are no grand narratives. In this, it offers more possibilities (Thompson 411).

Bernal, Werbner, Rai, and Thompson all characterize the diasporic public sphere in cyberspace as one marked by dissent, argument, and contestation. As Graham and Khosravi found, the explosion of Iranian-themed websites representing a vast array of opinions and insights means that what constitutes an Iranian virtual diaspora in cyberspace is becoming less inaccessible. They also note, "Some of the political programs found in cyberspace, such as demands for a restored monarchy in Iran or the political programs of the Iranian far left, if implemented, would certainly not lead to liberation and emancipation, but to new forms of repression and constraint" (222). Yet these sources also consistently invoke the frame of "constructing community" in their descriptions of the online practices of diasporic populations.

Definitions and uses of the term "diaspora" by scholars such as Cohen, Safran, and Hall are relevant to the Iranian case. All of these ideas about what the notion of "diaspora" offers have considerable utility for this study in terms of thinking about how this concept carries (1) certain oppositional identities which are in themselves heterogeneous and hybrid and (2) the potential for creating collectivities and mobilizing. These characteristics speak to a specific psychological stance associated with diaspora as being a force for new cultural expression. This "constructivist" approach to diaspora (Adamson) as an imagined community gets at internally constructed worlds and implies the possibility of agency and cultural production. It is not a cognitive leap, therefore, to see these purposeful efforts at constructing community among digital diasporas as a manifestation of kinship, the formation of a literal and metaphoric web of social relationships among dispersed kinswomen and kinsmen.

The Iranian diasporic public sphere is about not just interaction between diaspora members on the Internet, but interaction and engagement with other material structures such as voluntary associations and the localized receiving community. The public sphere also includes participation and mobilization in "extra-local" structures such as transnational political movements.

Using Virtual Ethnography to Locate Groups

The Internet is large. In order to uncover where a transnational community of Iranians exchanges true political discourse – and search for those points of consensus (or not), I employed virtual ethnography. According to Freidenberg, virtual ethnography consists of the following characteristics:

(1) the field site is comprised of internet users;
(2) the object of their study is their experience;
(3) the practices observed are virtual communications, that is, not face-to-face; and
(4) the purpose of virtual interaction is information exchange. (265)

Freidenberg contends that virtual ethnography is an excellent way to understand mobile and migrant populations (267). However, different virtual communities are not equal – each is characterized by different functions, aims, and discourses. For the virtual ethnography, I had to evolve and adjust tactics in order to pinpoint the location of the Iranian "digital diaspora," which, according to Laguerre, is

[a]n immigrant group or descendant of an immigrant population that uses IT connectivity to participate in virtual networks of contacts for a variety of political, economic, social, religious, and communicational purposes that, for the most part, may concern either the homeland, the host land, or both, including its own trajectory abroad. (50)

These diaspora-based online groups and Web forums enable the creation of a cyber community connecting dispersed populations and provide interactivity to members. Members use discussions forums to disseminate information, reinforce or recreate identity, and connect to and participate in homeland relationships, festivals, socio-economic development, and so on (Brinkerhoff 14). I sought groups that were not only overtly defined as Iranian but also oriented towards political topics or political discourse.

Exile Consciousness and Online Communities

No one category – immigrant, exile, diaspora, transnational, expatriate, hyphenated ethnic identity – fits the political experiences of Iranians outside of Iran. If the political activities and discussion Iranians outside of Iran engage in online are not neatly bounded by these categories, then how do we classify the actors? Understanding Iranian Americans' political engagement and participation as an immigrant politics, a transnational politics, a homeland politics, and a diaspora politics (cf. Ostergaard-Nielsen), sometimes simultaneously, goes a long way towards understanding their various political moments.

The contact point between all these moments is the presence of an exilic consciousness. Despite differences in the ability to return to Iran or immigration status, I would contend that it is the presence of an exile consciousness that permeates the discussions in online forums and constitutes community for Iranian immigrants. Exile can be externally mandated or internally imposed. This exile consciousness goes beyond mere nostalgia for something in the past, or as Brah words it, where home becomes a "mythic place of desire" (192). Exile discourse must deal with the continued problematic of multiple locations, where relationships are based not so much on shared origins (birth, nation, religion, gender) but on an attachment to a "common imaginary construction" (Naficy 2). Exile consciousness itself is longing for a home you cannot be a part of anymore as a full political member or transnational citizen. Regardless of individuals' ability to return, or whether they do make return visits, the fact is that on an emotional and psychological level, and in terms of their personal politics, they cannot be a part of the Iranian nation-state because they fundamentally disagree with Iran and the Islamic Republic. However, Naficy prods us not only to think of the dystopic experiences stemming from exile, but also to think of the utopic aspects of the exile experience, "driven by wanderlust and the desire for liberation and freedom" (6).

Facebook fosters the maintenance of transnational kin and friendship networks among Iranians in the diaspora. It is relatively easy to quickly log in to your Facebook account and see what your Iranian family members and compatriots living in Iran, Sweden, or the United States have been up to. For political or personal reasons, people may not be able to maintain face-to-face contact with their kin networks, but Facebook affords the possibility of a proxy kin network even with what would be considered "distant" family members through the scrolling of one's Facebook feed. For Iranian first- and second-generation Iranian ex-pats, distant family can still be present in their lives through enacting a virtual *meyhmooni*, albeit not in real time and edited to present the best possible image of daily life's happenings. The Iranian characteristic of trying to impress your fellow compatriots and maintain the best possible image can also be seen in one's Facebook posts and photos.

Another important aspect of Facebook is that for any individual user, it has the ability to convene friends, family, classmates, and work relations into one network. A user's post may be directed to only one aspect of life – say personal politics or something work-related – yet every other user ("friend") in that individual's network can see that post. For Iranians in the diaspora, and this is especially relevant to the second generation, transnational kin may get a glimpse of their life not offered or readily shared otherwise. It has the power to implicitly communicate more about ourselves and daily lives to distant kin than they otherwise would have known. For instance, if compatriots rely only on my father's infrequent trips to Iran bearing a couple of photos, I am only Hossein's youngest daughter with two children living in California. However, through my Facebook feed I am rendered more multidimensional and complex, and Iranian relations in my Facebook network have the opportunity to see me as more. This can be a positive thing, of course, because it fosters a deeper connection than bloodlines alone, yet on the negative side it means I am potentially under greater scrutiny and judgment as a representative of my family.

Facebook shares the dual purpose of connecting with friends and family in a purely social manner and being a site for disseminating news information about recent political events. Further, Facebook is multigenerational. I observed young 1.5- and second-generation Iranian migrants all the way into their early 70s who use Facebook, although younger users tend to dominate. Not all Iranian migrants use the Internet for politically oriented activities, yet all my informants noted a surge in online participation after the 2009 Iranian presidential election.

Because in the larger study that this is drawn from I used the analytical category of Iranian voluntary associations and organizations, I explicitly targeted Facebook groups to serve as the online corollary to on-the-ground Iranian ethnic voluntary associations. Preliminary research in 2010 revealed that among all the options in new media, Facebook and news sites (in Farsi or English) were used more than MySpace, Twitter, or even blogs. People tended to get their information from trusted news websites and then share and disseminate that information through Facebook. Though much has been made of the proliferation of blogs and alternative journalism in Iran (e.g., Kelly and Etling), my informants wanted to get their information from a trusted source. Facebook serves as a forum for activist groups as well as individuals to share and exchange information. For instance, groups might have their own website but also a Facebook presence that likely draws more viewers. Facebook users will post links from other sites and YouTube videos and other material. Therefore, Facebook might drive users to explore other political content that they otherwise might not have seen. For groups, Facebook offers users a more democratic forum for disseminating and responding to posts, where users from differing political leanings will "meet." For instance, Iranian American Facebook users with more conservative political ideologies might post on a group page for more left-leaning ideologies.

Easy ways to determine the popularity or currency of a group page are to look at the number of "likes" it has and to check the dates of the most recent posts. Based on this, I isolated two active "activist" groups on Facebook. I define activist as those Facebook groups that advocate for regime change, publish content against the Islamic Republic of Iran, and hold and organize rallies and other events, generally in opposition to the current government. I followed each group's Facebook wall content for a one-year period, September 2011–12.

Human Rights for Iran (HRI)[1] is based in San Diego, California, is composed of second-generation Iranian American university students, and uses Facebook to communicate with members, disseminate information, and coordinate events and meetings. I started following this group in late spring 2010 by attending some meetings on campus and being invited to join their "secret" Facebook group. Their secret group has 41 members, while their public group has 112 members. The public one has significantly more activity in the form of posts and comments, and I did not notice that the closed HRI group wall had significantly different content that might be labelled as secret. My sense was that

---

1 Names of organizations and informants are all pseudonyms.

the closed group was deemed secret primarily to restrict and control membership and not necessarily for content. The Facebook group's activity page reflects the fact that members' primary identity is as students. Lulls in Facebook activity mirror their student schedules, including breaks. I also saw a decline in this group's activity both virtually and on the ground as time passed after June 2009.

HRI's Facebook group page extends the reach and effort of their organization. For example, what is posted and discussed on Facebook is actually broader than what transpires at meetings, such as an effort to get members to "like" a page titled "4 Minutes to Prevent War with Iran" posted on their public Facebook wall on 2 May 2012 by the president of the organization. The president also announced that individual letters were going to be written during the next meeting and urged members to share the effort on Facebook, and he or she reminded members that "the cause is bigger than any one of us but together we will be the legs that [lift] the slumbering giant," meaning the hidden population of Iranian immigrants in the United States that would be opposed to a war with Iran. This particular initiative was first introduced online through the group's Facebook page and seemingly completed at a meeting. The most frequent kind of posts comes in the form of links to headlines and news stories from major journalistic outlets, trending topics like Iran banning university women from certain courses, the effects of US sanctions, and the three Americans who were hiking at a tourist spot near the Iran-Iraq border and were detained by the Iranian government over allegations of being spies before being released by 2011. Over the one year I tabulated posts, the news stories of interest included posts related to the anti-war on Iran movement (seven posts), the detrimental effect of US sanctions on the citizens of Iran (four posts), and the political situation in Syria (two posts), for example.

One of the more important functions of Facebook for HRI is to publicize their special events, such as movie screenings, rallies, and lectures. The political rallies posted on Facebook include a University of California system-wide rally for Syria, a candlelight vigil for victims of hate crimes, a Los Angeles event sponsored by Democracy in Action which called for "No war on Iran" and against foreign intervention in Iran, and an event in Los Angeles about the oppression of the Bahá'í faith.

Posting links to online petitions for certain causes, such as the anti-war with Iran effort, was yet another use of HRI's Facebook wall. Online petitions compete in an environment on HRI's Facebook page already crowded with news information and calls for action, such as YouTube videos, political rallies, and documentary movie screenings. In addition, my informant Niki specifically counted signing and forwarding online petitions among her political activities.

Another group, Iranian American Youth (IAY), was initially local to the Washington, DC, region but by 2012 aimed for a national reach. This group is public and had 1,819 members as of October 2012. IAY is primarily a Facebook group with a concomitant website. Their Facebook content consists of news stories posted from other outlets, with no ensuing group discussion. IAY was organized out of the support for Iranian protesters after the June 2009 election. Around June 2009, IAY did organize support protests but no other major events since that time. Iranian American Youth has been virtually inactive since 2010.

Analysing the political engagement of particular associations of Iranian immigrants through the groups feature on Facebook is relatively one-dimensional, yet Facebook can serve as a rough barometer for the political impulses of diasporic Iranians. There was an explosion of Facebook groups during the height of the 2009 Iranian election crisis and protests, including some devoted to a young woman protester named Neda Agha Soltani, killed by gunfire during one of the Tehran protests. The image of Neda's face lying in the streets quickly became an icon of the Green Movement's protests and circulated around the world.

The term "Facebook activism" has made its way into the Urban Dictionary, albeit in a sarcastic manner, denoting "[t]he illusion of dedication to a cause through no-commitment awareness groups" (n.pag.). There is an increasing sense that Facebook activism is a poor indicator of actual commitment to causes. Yet there is also a proliferation of how-to guides and manuals for organizers to take advantage of Facebook's widespread use around the globe, no cost for organizing, and ease of use of multimedia tools. The manual *A DigiActive Introduction to Facebook Activism* (Schultz) is an example of such an effort. The author of the guide views Facebook as a great way to increase awareness and mobilize people. However, Mary Joyce, co-founder of DigiActive.org, commented on the low bar of entry for Facebook groups and activism. In a 2009 article, Joyce explains, "Maybe a maximum of 5 percent are going to take action, and maybe it's closer to 1 percent … In most cases of Facebook groups, members do nothing. I haven't yet seen a case where the Facebook group has led to a sustained movement" (Hesse, n.pag.). Whereas the great majority of Facebook activism involves little effort on the part of the user or does not translate to sustained movement, this is not always the case.

For one of my informants, Mohsen, his Facebook activism comes not from clicking "like" on particular groups but through news and information-related posts. These posts on his Facebook wall are real-time connections to events happening in Iran, and he sees the impact of his posts when friends and relatives in Iran thank him for disseminating news and information that they might otherwise not be aware of. He does concede that these efforts may be small, but he sees it as contributing overall to regime change in Iran. Mohsen argues,

I can say that because I was worried, I always loved to [see] something happened in Iran that change the regime. But I was never doing anything about it. And still I'm not doing anything. But the only thing I can do is post these things in on Facebook. You know, those days on Facebook some people were changing their names to I don't know, Mohsen Irani, Mohsen Tehrani, they don't give their pictures, don't give their information. Mine I didn't change it. From the first day still is the same thing. My name and family. And I'm posting. The only thing I am doing is the posting the news I'm receiving. Posting there to, for the people who are interested in it. And, just wish that it helps and it brings another person to think and to act. I am not there. If I was there maybe I was more active. But here whatever I can do is that. And yes I get more active in political issues when I, when this happens in Iran. Give me some hope that something can happen in Iran, regime can change in Iran. It's not a dream. It can be done. Maybe takes time, but [it] can be done.

Facebook as a social networking site is one arena where political activity takes place – on the individual level through Mohsen, who sees Facebook as a useful tool to advance regime change in Iran, and on the group level through HRI. While Facebook is a useful medium to disseminate information and bring together like-minded people to participate in political and social causes, and to perhaps "introduce" topics for consideration by users and members, it was not used for any *engaged* discussion of political topics. I had hypothesized that Facebook groups operated similarly to on-the-ground Iranian ethnic associations and basically functioned as a parallel to kin/family groups, except for political issues. The groups function that I was so keen to study as a rich resource of political dialogue and engagement did not bear out. In other words, Facebook was used as a tool of local forms of civic engagement but not as a locale for transnational political discourse. I found that more politically oriented discussion and interaction actually takes place on a forum like Iranian.com.

On Iranian.com

Iranian.com (IC) is a self-described community site for the Iranian diaspora, for "the Iranian expatriates who care about their identity, culture, music, history, politics, literature and each other, as well as friends and family living in Iran" (n.pag.). Its byline is "nothing is sacred" with an illustration of a goldfish, a symbol of the Persian New Year. This motto, according to the "About Us" page, "reflects our view that religious, political, cultural, or commercial considerations should not prevent the publication of any material." Founded by publisher and journalist Jahanshah Javid, it has been in existence since July 1995. In the first year the site was updated every two months, as it took that long to write and publish material and develop new content. According to Javid, by 1997 it was updated every day as its readership grew and people submitted articles. In 2007, Javid teamed up with a group of private investors from Northern California who funded the site to go interactive (Karami, n.pag.). As of 2 July 2012 the site had 7,830 registered members, most of whom are from North America, Europe, and Australia, in that order. Iranian.com has a limited number of viewers from Iran because it is blocked by most Internet providers ("Frequently Asked Questions").

As with Facebook, IC users must register as members to post content such as articles or to comment on others' content. Registered users are referred to as bloggers, and each blogger has their own page. Most often, usernames are pseudonyms that may reflect their politics or origin. There are many more registered users who commented once or twice and many fewer users who post regular content for their blog as an initiating point for discussion and comment regularly on others' posts.

The site has a busy interface with various top and side banners, such as an "Iranian of the Day" section usually featuring political prisoners. The middle area contains headings and links to individual blog posts, showing recent comments, most commented posts, and most viewed posts. Though the top tabs feature topics like "Music," "Photos," "Arts & Lit," "Life," and "Football," in reality the news and blog sections of the site are the most active and commented upon. The vast majority of the site's content, posted by its members, is news, comments, and opinions of a political nature.

Broadly, the major differences between Facebook and Iranian.com are that Facebook is based on an individual user employing their true identity and profile information, "friending" others, and thus creating a network. The groups function in Facebook is actually secondary. IC has users with no associated profile information and allows avatars and multiple identities. Thus IC members are mostly anonymous unless they choose to use their real identity, and most do not. Users can have their own blog on IC and post original articles, as well as comment and post on others' blogs. However, there is no "network" between members. IC is also editorialized and, though not censored for content, is monitored to ensure that a degree of civility is maintained.

I classify Iranian.com as a diaspora website for two reasons – first for the kind of discourse and exchanges encountered there (elaborated on later), and second for how the amount and frequency of user posts correlates with major news events for the Iranian diaspora. To determine how closely the activity on Iranian.com is linked to Iranian news, I used the site archives to tabulate the number of posts by month and year. Site archives are not available before July 2007.

As might be predicted, the summer of 2009, during the height of the post-election drama in Iran, shows the highest number of posts. July (1,859), June (1,724), and August (1,580) of 2009 constitute the three highest monthly totals for posts in the five-year period for which archives are available, in that order. This is followed by December 2009 and September 2009. Excluding 2007, when the site was first going interactive (thus the number of posts are artificially low), June, July, August, and December (in that order) have the highest monthly averages from 2008 to 2010. Iranian.com is more or less an accurate barometer for the "pulse" of the Iranian diaspora.

Dissension and Disagreement

Iranian.com is a bilingual Farsi/English diaspora website. There is a small portion of users who post exclusively in Farsi (and that number has decreased since the site's launch), but the vast majority of its content is posted in English, thus making it readily accessible to members of the second generation. The focus of this analysis is on English-language content. Despite the other topic headings, news and discussion of a political nature dominate the site's content. Through following political talk and politically oriented discussion on IC, I found first that IC provided a forum for the safe expression of a diverse range of political, ethnic, and religious difference not found in face-to-face interactions among migrant Iranians. Yet, within this, the discourse on IC is much more strident and divisive than is heard and experienced in face-to-face life. Therefore, there is conflict, disagreement, dissension, and certainly at times uncivil discourse, yet differing ideologies are expressed in an environment without violence.

Whereas my informants in San Diego assured me that differences in political ideologies were never a big issue in social situations, at *mey-hmooni*, or during monthly meetings of Iranian ethnic organizations, IC users use the forum to seriously discuss their position and try to convince others of the rightness of their view. IC is certainly a community, with dissent and disagreement at its core. After a bit of time you get to know the personalities involved. It is instructive to hear some news about Iran in mainstream media and then to go check on Iranian. com to gauge how the news is being received and analysed by the Iranian diasporic community. Online political discourse here, besides a propensity to be insulting or even bigoted at times, consists of ardent declarations of specific political programs and ideologies that turn into animated debates in many instances. The commentary by user Esfand Aashena posted 14 September 2012 is an example. This user takes issue with others insulting Islam and does so by pointing out the ineffectiveness of their insults in changing the regime:

> So I turn to those who insult Islam routinely on this website and refer to Muslims (directly or indirectly) as savages and Muhammad as a pedophile and so on, dirt on your heads! NONE of your insults on this website has ever been able to incite anything in Iran! ALL of the protests in Iran have been because of insults of non-Iranians!

Similarly, this comment by user Darius Kadiver clearly articulates his pro-monarchy stance on 4 September 2012:

> I'm a Monarchist Period not a Celebrity Seeking "FB Pahlavist" Fashion Victim ... And have been So From DAY ONE!
>
> I didn't wait to join the Pahlavist bandwagon (if any ?) 3 Decades later to express my Pro Monarchist views, because I believe in them, and did so even before contributing to this website some 12 years ago ...
>
> So no need to be grandstanding when lecturing me on an era you clearly never lived under.
>
> Dunno how old you are or from which planet you are speaking from (I'm sure Armstrong could tell ... Sorry I can't ...) but clearly you like your like minds are talking on a period you either were too young to remember

or are one of those Post Revolution generation Schizos who was brainwashed into believing any crap fed to them by your own intellectually bankrupt ex revolutionary parents added to the IRI [Islamic Republic of Iran] propaganda machine.

Then I guess probably following the Post Election Crackdown, you suddenly realized that your Joon Jooy Republic was not such a "Behesht" your former "Presidenteh Mahboub" Khatami claimed it to be but rather the unreformable shit hole it always was from Day On [sic].

Other times, I witnessed users hurling accusations at others about being "agents of the Islamic Republic of Iran." It is difficult, however, to gauge if all the ideas presented on IC have equal weight and measure among the Iranian diaspora at large. Because users can hide behind almost total anonymity if they choose to, without even any knowledge about where they are posting from, and because this political discourse is isolated from other phenomena related to life course and identity, it is difficult to get a sense of how political talk on IC relates to other types of political practice on the part of users. In other words, do the most active users of IC engage in other types of political or civic behaviours? Another real possibility is that there is not necessarily a 1:1 correlation between user identities and actual users. One accusation bandied about is about users who have multiple IDs and post similar inflammatory content under different user names, thus seemingly driving up the popularity of their views.

Reinforcing Discourses – Creating a Political Community

In some cases, evidence from analysing the discourse in online communities reinforces what I found through conventional ethnography. For example, I discovered a couple of concepts about civic/political participation and its broader meaning that informants had in common with Iranian social media users. It is noteworthy that applying the analytical category of diaspora groups for both on-the-ground Iranian-based ethnic organizations and online groups yielded similar types of political discourses. The first shared discourse is a theme centred on criticizing the lack of political participation in the US system among Iranians, Iranians "taking advantage" of US society, and the impetus to give back to society. The second shared discourse is about Iranians learning how to live in a democracy by first practising values like tolerance and equality through participating in such organizations.

Lack of Political Participation in the United States and Lack of Community

As evidence of the currency of this particular discourse, I offer the following case study from Iranian.com. "Anonymous Observer" (AO) was the author of a blog article titled "The Good American: Why Iranian Americans Will Never Advance" on 3 September 2012. The gist of the article was a criticism levelled at fellow Iranian Americans for being disconnected from American politics and politically unaware. The starting point was a discussion on IC about Texas Representative and Republican presidential candidate Ron Paul. Anonymous Observer in his/her blog post details why a Ron Paul presidency would be disastrous for Iranian Americans. AO's tone is one of strong disapproval and condescension towards Iranian Americans lacking knowledge of and involvement with US politics:

> But none of the above facts, and Ron Paul's scary vision for America seems to bother his Iranian supporters. They have no issues with America being run by a racist and for their children losing their citizenship. There [sic] only concern is the "old country" and the ability of mullahs to have access to the latest technology, free trade and lots of lots of petrodollars. And this, my friends, is the single most important reason why the Iranian American community will never advance. Iranian Americans are too attached to the old country. The umbilical cord has not yet been cut. Their main focus should be the United States, but it's not. It's Iran.

Then AO chides Iranian Americans for their lack of loyalty to the United States and goes on to blame Iranian people themselves for the 1979 "devolution," as AO calls it:

> Remember: your American passport is not only for ease of travel to Dubai on your way to Iran for your annual chelo-kabab[2] feasts. You took a loyalty oath to this country that should not be dismissed as "alaki."[3] And you're not "zerang"[4] for becoming U.S. citizens. This country trusted you. Don't betray that trust. Also, remember this: you don't know more than an average American just because you had a "revolution" back in 1979. Look at what

---

2 Skewered beef grilled over hot coals
3 Fake
4 Clever

that fiasco did to your homeland. If anything, your 1979 devolution is proof positive that you know absolutely nothing about running a country.

The blog article generated 149 total comments, which is on the higher end for the site in the preceding months (but nowhere near the highs of 800 to over 1,600 comments generated by a few discussion threads in 2009). Among the 22 users who posted comments, only a handful (six) agreed with AO's overall claim. One of those, Hafez for Beginners, blamed the issue of lacking political representation on the community's inability to resolve differences. Others called AO a traitor and overly insulting and accused AO of "generating propaganda." A good portion of the comments devolved into arguments on specific points between posters that were divided along political fault lines – apologists for the Islamic Republic of Iran, also called "Islamists" by others; those who critique the American "empire" yet are adamantly against the IRI; and others proudly refusing to engage with anything American. Frashogar, another Iranian.com user, posted a response to AO's article accusing AO of engaging in hate speech:

> Ignore all you like. It doesn't change the fact that you and your friends on this website are engaging in clearly identifiable and legally definable **hate-speech** [emphasis in original] against the entire Iranian community, where you and your friends here have even crossed lines into incitement and advocacy of murder countless times. Wrap yourself all you like under transparent criticisms and the flag of the American Empire. When it walks like a duck, quacks like a quack, by golly it is a duck.

In IC, the discourse about a perceived lack of community surfaces whenever discussions between users begin to break down along religious and political fault lines. Phrases like "whoring for Israel and Bahá'í's" and other similarly bigoted and intolerant ideologies can occur in the same post as accusations of being anti-Iranian and are obviously meant to incite division. Sometimes such users are reprimanded by other members in a more reasonable tone, as in the admonition Hafez for Beginners posted 6 September 2012:

> No community will survive like that and have a real voice, if they can't put up with differences.
> Better luck to the next generation – I guess – and I do think those born in the US don't have the baggage and can learn to listen and put up with differences better.

Or, by user Mohammad Ala, in a comment from 12 November 2011 titled "No social responsibility," responding to a series of increasingly insulting posts being exchanged between members:

> Iranians come from different family backgrounds, they grow up in different parts of Iran, most of them pretend to be religious. Failure of our community demonstrates why we have not been successful (as a group). Many times I have observed ISP members resort to personal attacks to make a point. Is it necessary to criticize someone's picture, name, or his/her title? I agree this is not limited to Iranians. As I wrote before, we are good in three so-called Rs[5] (math, reading, and writing) but fail big in the fourth R (social responsibility).

Learning Democracy

The second noteworthy theme I observed was a discourse about the need for Iranians to "learn democracy." Specifically, metaphors of the personal were inscribed onto the national body politic. There was the notion that psychosocial growth and development in personal attributes such as "being able to tolerate differences" signalled fundamental change for the larger community and nation. Iranian diaspora online forums thus may act as catalysts for such changes. A metamorphosis in one's own ability to tolerate different viewpoints leads to greater changes within the larger society. Therefore, the only way the Islamic Republic of Iran can move forward is with the idea of "change from within" – in other words, bottom-up change from individuals learning democratic ideals and notions of freedom. There is also the awareness that it is through participation in the institutions of US civil society, including a diasporic public sphere, that Iranians have learned about democratic ideals such as tolerance and respect for free and open speech.

In some instances it is when discussion breaks down between individual members into extreme negativity and name-calling that members might make a general statement that Iranians still "have a long way to go" and that tolerating different viewpoints is a "habit Iranians

---

5 "The three Rs" is an idiomatic expression that refers to "reading, writing, and 'rithmetic" (arithmetic).

never learned," as in the following post from Anonymous Observer on 4 September 2012:

And to add – Iranians (especially the older generation) are not used to a civilized political discourse. Look at what I have been called just in this blog – by people who don't know a single thing about me – for offering a point of view!! We have a long way to go in learning how to deal with differing points of view. Hopefully, the new generation will fare better than these obsolete fossils.

A user named Mehrban on 12 October 2012 credits Iranian.com with helping him to learn how to tolerate different viewpoints and with being the one and only place where Iranians can freely express themselves, unlike in social networking sites where participants need to present their "best face to the world."

There has been one feature to IC and only one feature that sets it apart from other sites, people (Iranians) with different points of view could learn to discuss with one another, it is a habit we have never developed, I personally have learned it here (if I have learned it).

Being from a hierarchical society we have little room for opposing ideas and do not have a capacity to even reason through our own ideas because often we don't even know why we think what we think. Esfand jaan,[6] I don't expect you to understand the importance of this feature, as you have consistently opted to express yourself in leechaar[7] as opposed to a reasonable discourse.

This is not a social networking site, where you need to present your best face to the world. It is THE ONE AND ONLY [emphasis in original] place where Iranians can freely express themselves and learn to tolerate each other.

My conversation here is mostly with Mr. Amin and I am sorry that it is taking place in your blog. I think it is my responsibility as a reader and a contributor of this site to emphasize to with Mr. Amin what it is that actually makes his site not only unique but also worthwhile.

BTW, Esfand jaan, unlike you I don't find most blogs worthless, by being here I have learned a whole lot and cherish my daily visits to this site.

---

6 Esfand was the author of the original blog article that Mehrban is responding to. "Jaan" is a common term of endearment meaning "dear."

7 Closely related to the idea of provoking.

Now go ahead and try to degrade my heartfelt insistence of something extremely important to haggling or whatever your aggressive leechaar lexicon prompts you to do. You cannot (yet) delete my reasonable comment :).

The larger context of this comment was the question of whether the authors of blog articles should be able to moderate their own content and delete responses as they wished. Mehrban felt that deleting others' comments, no matter how offensive to the author, was akin to shutting down true and open discourse. Another response post by Mehrban from 10 October 2012 pertains to a discussion about the potential of Iranian.com being shut down and offers the idea that discourse on IC should aim to be a model of what open discourse in society writ large should look like:

Dear Mr. Amin, I am not sure if a blog is someone's home and even if it is someone's home it is a home whose door has been opened to the public.

What does it mean if someone is interested in presenting their ideas to the public but not interested in hearing opposing views? How can you stop your site from becoming a propaganda scene, or a hollow self-aggrandizing venue without any challenge (checks and balances) from other readers/bloggers?

Shutting out opposing views is what we (Iranians) are very good at and needless to say it is the root of many of the ills in our society. At IC we have learned that it would not kill us to hear something that we don't agree with and in time we each have developed ways of dealing with it in a way that has not required killing (deleting) each other.

Further, Mehrban feels that individual posters need the "checks and balances" provided by other users commenting on their ideas to keep the site from devolving into propaganda.

IC convenes a large number of Iranians in the diaspora (more than any other site purportedly), and it is a politically and ethnically heterogeneous group. There also exists more diversity in religious ideology and even ethnic diversity in IC than encountered in local populations of Iranian migrants. As such, it establishes a political community where competing ideologies and frames of reference are presented and considered by its members.

Debating and discussing homeland, host country, or transnational politics through online forums can be seen as a sign of immigrant political agency. It is through participation in these virtual venues that

Iranian migrants learn participatory democracy, including the idea of "social learning about civic engagement" (Brettell and Reed-Danahay 198), and finally, enact democratic values (Brinkerhoff 2) within the institutional contexts of nation-states. That the discourse on IC in particular more often turns hostile and negative is seemingly alien to the very democracy it attempts to enact. As Bernal found, it is actually through conflict and encountering difference, not sameness and consensus, that identities are constructed and community constituted (669).

If the ability to express differences through the venues in which political discourse takes place among Iranian immigrants marks a democratic ideal, then we must accept the irony that it is through this expression, no matter how filled with disagreement, that the diaspora enacts democracy and creates transnational civil society and political "community," where community signifies the existence of a diversity and multiplicity of voices. I believe the presence of shared discourses indicates the formation of a distinct Iranian American civil society. An Iranian American civil society is distinct because it references the unique political history of Iran and Iranian categories of self-responsibility, personhood, and the components of *communitas*.

Conclusion

I began this study with the notion that Iranian online groups were the virtual corollary to on-the-ground Iranian voluntary associations – basically a forum where people who had similar political interests could come together. Further, in the case of exile or displacement of migrants, actual kinship ties may be severed or otherwise rendered ineffective; therefore, I surmised fictional kinship groups (online and offline) are created to make up for this.

Quick media and new modes of online engagement alter social organization, attitudes and personalities, and social relations in general. It is largely through the activity of various voluntary groups that migrants express their personalities, as well as acquire and broadcast status. As a sense of shared "territory" becomes less and less a source of solidarity among migrants, more differentiated sets of interest units are created. The Internet certainly is a forum for engaging politically among migrants and creating discursive virtual kin networks based on the exchange of political talk. Not all online migrant political activities are the same, however.

Mohsen's Facebook activism and HRI's use of Facebook for activism and organizational coordination evidence at least two different uses of Facebook for political participation and engagement. It seems networks like Facebook can lead migrant users to new insights about their own role in enacting democratic values because these venues are (a) not anonymous and (b) convene friends, family, and work relations from all aspects of life, resulting in a more engaged, and potentially more vitriolic, back-and-forth dialogue.

Facebook differs significantly from the way Iranian.com creates a political community composed of a diversity and multiplicity of voices engaging in political discourse. Iranian.com brings together multiple generations (older migrants, millennials, and second generations) and differing political factions and ideologies into one arena. One reason behind this might be that IC users do not know each other in "real" life and are thus anonymized.

Yet Facebook offers a less schizophrenic engagement because most parts of one's life are integrated and on display. It provides a "consistency and integrity" of personality that other diaspora websites may not. Among Iranians at least, the network of Facebook friends is genealogical. There is subtle pressure to "friend" any Iranian relation who requests to be included in your network. However, with the groups function, and to the same degree the pages liked function, these ties are selectively chosen and based on common interests and values. The danger is that our Facebook networks are only composed of similar "interest units" – basically other people who feel and act along the same political spectrum that we do. The advantage Iranian.com offers in this regard is that the "territorial unit" of Iran as a nation-state is the sole common interest creating solidarity, but within this there are many kinds of opinionated factions presenting themselves, which users must negotiate. And at the same time, the relationships engendered by Facebook networks potentially allow for greater opportunities for physical world engagement, while Iranian.com's politically engaged talk remains a more limited kind of mobilization. As our collective use of these new media continues to evolve, we need to stay attuned to the differences in the ways they articulate with our political selves, as well as who we choose as kin.

The current geopolitical relationship between Iran and the United States structures possibilities of action and prevents most types of transnational participation by Iranians living abroad. When there are hostile relationships between host and sending countries, "the use, form, and mobilization of the connections linking here and there are contingent outcomes subject to multiple *political* constraints" (Waldinger and Fitzgerald 1178–9, emphasis in original). Among Iranian immigrants there is the additional factor of mixed migration status affecting and structuring possibilities of political action. Iranian political refugees are absolutely prevented from returning to Iran in their lifetime. Therefore, what overall purpose does immigrant political agency as expressed through online groups serve, beyond constituting community? No diaspora could overthrow a repressive regime in its homeland (Sheffer 215). Yet their trans-state networks, such as using Facebook to disseminate news and information to friends and family in Iran, and in other cases through the global network of Bahá'í's lobbying in host counties, or discussing the possibilities presented by the Green Movement on a forum like Iranian.com, "diasporans can foment internal instability and tensions in their homelands" (Sheffer 215). In this age of instantaneous communication, these kinds of quick media enact kin networks through the exchange of political discourse and ideas, and thus build an immigrant civil society in cyberspace, while helping the overall cause of regime change in Iran.

WORKS CITED

Adamson, Fiona. "Mobilizing for the Transformation of Home: Politicized Identities and Transnational Practices." *New Approaches to Migration: Transnational Communities and the Transformation of Home*. Ed. Nadje Al-Ali and Khalid Khoser. New York: Routledge, 2002. 155–68. Print.

"About Us." *Iranian.com*. Iranian LLC, n.d. Web. 7 Oct. 2012.

Amini, Soheyl. "Information and Communication Technologies and the Changing Identities of Iranians in the 21st Century." Middle Eastern Studies Association Conference. Manchester Grand Hyatt, San Diego, CA. 18–21 Nov. 2010. Conference Paper.

Bernal, Victoria. "Eritrea On-Line: Diaspora, Cyberspace, and the Public Sphere." *American Ethnologist* 32.4 (2005): 660–75. Web. 16 Nov. 2010.

Brah, Avtar. *Cartographies of Diaspora: Contesting Identities*. New York: Routledge, 1996. Print.

Brettell, Caroline B., and Deborah Reed-Danahay. *Civic Engagements: The Citizenship Practices of Indian & Vietnamese Immigrants*. Stanford: Stanford University Press, 2012. Print.

Brinkerhoff, Jennifer M. *Digital Diasporas: Identity and Transnational Engagement*. New York: Cambridge University Press, 2009. Print.

Cohen, Robin. *Global Diasporas: An Introduction*. Seattle: University of Washington, 1997. Print.

"Facebook Activism." *Urban Dictionary*. Urban Dictionary, 7 May 2009. Web. 2 Oct. 2012. <http://www.urbandictionary.com/define.php?term=Facebook%20 Activism>.

Fraser, Nancy. "Rethinking the Public Sphere: A Contribution to the Critique of Actually Existing Democracy." *Social Text* 25/26 (1990): 56–80. JSTOR. Web. 13 Feb. 2012.

Freidenberg, Judith. "Researching Global Spaces Ethnographically: Queries on Methods for the Study of Virtual Populations." *Human Organization* 70.3 (2011): 265–78. Print.

"Frequently Asked Questions." *Iranian.com*. Iranian LLC, n.d. Web. 5 Oct. 2012.

Graham, Mark, and Shahram Khosravi. "Re-Ordering Public and Private in Iranian Cyberspace: Identity, Politics and Mobilization." *Identities: Global Studies in Culture and Power* 9.2 (2002): 219–46. Print.

Habermas, Jürgen. *The Structural Transformation of the Public Sphere: An Inquiry into a Category of Bourgeois Society*. Trans. Thomas Burger. Cambridge, MA: MIT Press, 1989. Print.

Hall, Stuart. "Cultural Identity and Diaspora." *Identity: Community, Culture, Difference*. Ed. Jonathan Rutherford. London: Lawrence and Wishart, 1990. 222–37. Print.

Hesse, Monica. "Facebook Activism: Lots of Clicks, but Little Sticks." *Washington Post* 2 July 2009. Web. 2 Oct. 2012.

Karami, Arash. "Q&A: Jahanshah Javid: From Iranian.com to Iroon.com." *Frontline: Bureau* 28 Aug. 2012. Web. 24 Sept. 2012.

Kelly, John, and Bruce Etling. *Mapping Iran's Online Public: Politics and Culture in the Persian Blogosphere*. Berkman Center for Internet and Society. Harvard University, 5 Apr. 2008. Web. 4 Aug. 2010.

Laguerre, Michel S. "Digital Diaspora: Definition and Models." *Diasporas in the New Media Age: Identity, Politics and Community*. Ed. Adoni Alonso and Pedro J. Oiarzabal. Las Vegas: University of Nevada Press, 2010. 49–64. Print.

Naficy, Hamid. *The Making of Exile Cultures: Iranian Television in Los Angeles*. Minneapolis: University of Minnesota Press, 1993. Print.

Ostergaard-Nielsen, Eva. "The Politics of Migrants' Transnational Political Practices." *International Migration Review* 37.3 (2003): 760–86. JSTOR. Web. 13 Dec. 2010.

Rahimi, Babak, and Elham Gheytanchi. "Iran's Reformists and Activists: Internet Exploiters." *Middle East Policy Council* 2008. Web. 10 Sept. 2012.

Rai, Amit S. "India On-Line: Electronic Bulletin Boards and the Construction of a Diasporic Hindu Identity." *Diaspora: A Journal of Transnational Studies* 4.1 (1995): 31–57. Project Muse. Web. 18 Oct. 2012.

Rumbaut, Rubén G. "Ages, Life Stages, and Generational Cohorts: Decomposing the Immigrant First and Second Generations in the United States." *International Migration Review* 38.3 (2004): 1160–205. JSTOR. Web. 4 May 2010.

Safran, William. "Diasporas in Modern Societies: Myths of Homeland and Return." *Diaspora: A Journal of Transnational Studies* 1.1 (1991): 83–99. Project Muse. Web. 12 Aug. 2008.

Schultz, Dan. *A DigiActive Introduction to Facebook Activism.* DigiActive, 2008. Web. 18 July 2010.

Sheffer, Gabriel. *Diaspora Politics.* New York: Cambridge University Press, 2003. Print.

Thompson, Kenneth. "Border Crossings and Diasporic Identities: Media Use and Leisure Practices of an Ethnic Minority." *Qualitative Sociology* 25.3 (2002): 409–18. Web. 18 Oct. 2012.

Waldinger, Roger, and David Fitzgerald. "Transnationalism in Question." *American Journal of Sociology* 109.5 (2004): 1177–95. JSTOR. Web. 18 Sept. 2012.

Werbner, Pnina. "Diasporic Political Imaginaries: A Sphere of Freedom or a Sphere of Illusion?" *Communal/Plural* 6.1 (1998): 11–31. Web. 16 Nov. 2010.

3 Negotiating Womanhood and South Asian Nationalisms: Blurring Borders and Identities in Social Media

APARAJITA DE AND SHEKH MOINUDDIN

What is the difference between Indians and Pakistanis? The answer is uncomplicated: There is no difference. We are the same people, with similar personality strengths, and parallel collective weaknesses. Why then have the two nations moved along such dramatically different arcs in the six decades of their existence?

M.J. Akbar[1]

The Nation in Everyday Social Media: Remapping India and Pakistan

When a leading journalist and intellectual like M.J. Akbar, who also happens to be an Indian Muslim, suggests there is no difference between India and Pakistan, a closer look is required. Any imagination of India and Pakistan is wholly informed by difference – the two separate nation-states are articulated and embodied by the borders that are largely assumed to create lines of division that fragment people and spaces on the basis of religion and other differentiations. As Homi Bhabha has argued, the dominant master narrative seems to be the "language of those who write of it" (1); the master narrative thrives on political ideologies. In other words, it is those representational practices which define, legitimize, and valorize a particular idea of the nation-state and notions of nationalism (During). The borders between India and Pakistan are so deeply entrenched that all alternative or

---

1 M.J. Akbar is a well-known Indian Muslim journalist and author. He was formerly the editorial director of *India Today*, a leading English newsmagazine.

counter-monolithic narratives on either side of the border are viewed with great doubt and scepticism. However, it would not be wrong to conclude that both nations have been and are experiencing virulent forms of nationalism, particularly in the context of rising fundamentalist and fascist ideologies (Varshney; Yousaf; Lankala). This chapter aims to explore a broader landscape of shifting identity markers through an analysis of the specific construction of a South Asian female tennis star, Sania Mirza. Before looking at the specificities of Mirza's positioning, however, it is important to explore some of the history of the region, as well as the digital backdrop upon which connections between Indians and Pakistanis occur.

Background

India and Pakistan are two neighbouring modern nation-states that were created in 1947 by British colonial powers on the basis of religion. The partition was a violent and traumatic moment marked with mass exodus and exchange of people from both sides of the border. The inheritance of this violent past culminates today in discourses about the highly divisive and bifurcated identities of both India and Pakistan. The subsequent wars of 1965, 1971, and 1999, constant cross-border skirmishes, the contentious issues of Kashmir and the highly politicized matter of Ram Janmabhumi (the birthplace of Ram), the Babri-Masjid conflict in 1992, and terrorist attacks in India have kept alive the identities of the two nation-states and hardened the borders of the two. All of these incidents created a highly visible public discourse of sealed borders, of absolute lines of separation that close virtually all forms of communication and encounters.

Despite this context of divisiveness and rancour, at the present moment, the master narrative of Indian-Pakistani relations seems poised to change. Two examples illustrate this shift. First, in 2013, *Daily Times of Pakistan*, a Pakistani English-language daily, ran a piece on the 107th anniversary of the birth of Shaheed Bhagat Singh, a freedom fighter against British imperialism. Although Bhagat Singh is identified as an Indian freedom fighter, many Pakistani now consider him to be Pakistani because he was born in present-day Faisalabad, Pakistan, and he was hanged by the then British government in Shadman Chowk, Lahore, in Pakistan. The celebration was marked by

participants holding banners and placards in his memory and distributed pamphlets containing a short biography of Bhagat Singh among the citizens.

Later, senior participants of the gathering cut a cake to mark the 107th birthday of the socialist. A candlelight vigil was arranged on the occasion to remember Bhagat Singh's exemplary contributions for the oppressed. The speakers paying tributes to Singh said he was the hero of the people of the Indian Subcontinent who struggled against the British to free the lands from the foreign occupants ... speakers also pledged to carry on the struggle against the oppression and imperialism to realise Singh's dream. ("107th Birthday")

The report also talked of Bhagat Singh's followers wanting to rename Shadman Chowk, where Bhagat Singh was hanged, as Bhagat Singh Chowk. This example situated Bhagat Singh as a leader and fighter for Indian and Pakistani people collectively, highlighting commonalities rather than divisions between the nations.

Shortly afterwards, in the same year, India's largest circulating English daily, *Times of India*, reported on a Bollywood film *War Chhod na Yaar* (Leave the war Dear) being made by debutant director Faraz Haider, which "showcase[s] the relations between the two countries in a light manner" ("War Chhod Na Yaar" 8) and is being hailed as India's first war comedy. The director acknowledged that most of the films made on Indo-Pak issues have shown Pakistan in a bad light, given that anti-Pakistan sentiment has been the common factor in all the films on Indo-Pak wars, and sought to provide an alternative view.

These two newspaper reports are indicative of the slow but steady emergence of a parallel counter-narrative that talks of blurred boundaries and of shared cultures, religions, and traditions, shared spaces, and a shared British colonial history. Seeing the two nations beyond being binary opposites pitched against one another conveys the borderlessness many Indians and Pakistanis experience through the use of quick media. The particular emergence of social media not only rendered the borders porous but also enabled cross-border communication, which was previously not possible to the same extent. Our chapter closely examines the emergence of this interface and how transnational identities are created and negotiated within the context of these blurring borders. Set in this particular dominant master narrative of India and Pakistan, the essay looks into how everyday conversations over quick media forge kinship ties and create new transnational identities and a larger South Asian identity on the basis of these shared cultures and histories.

Looking at Sites of Connection

This chapter takes on a qualitative examination of the conversations taking place over social networking sites, websites, blogs, and chat logs like Patheos, Gupsup, Merinews, and Facebook pages. The material under discussion focuses on tennis player Sania Mirza. Our analysis suggests ambiguities in the dominant master narrative and the emergence of a counter-narrative. We analyse continuous blog threads in order to examine the dialogue that people engage in through the use of quick media. These threads are part of public Web pages which are not membership based and do not feature moderated conversations.[2] These everyday conversations are mostly facilitated through various India-Pakistan Friendship groups and other groups on social media networking sites focusing on a range of shared interests, from cricket, music, poetry, and films to politics. In our quotes from these sites, we do not clean up language idiosyncrasies because they capture local flavour and nuances. Most of these sites highlight a continual reproduction and reconstruction of the histories of two nations. For instance, the introduction of the Facebook group The Indo-Pak Bangladesh Friendship Forum states,

> India and Pakistan came into existence in August 1947, out of pre-partitioned India, as a result of British colonialism handing over power to political forces of two newly created countries … In spite of bitterness, the people of both countries were not denied frequent visits without much hindrance till 1965 war, which changed the scenario between two countries. Later 1971 war, creation of Bangladesh created further bitterness. Whatever manner the three countries – India, Pakistan and Bangladesh came into existence, the fact is that people of all three are bound with strong cultural ties with each other, on one side two divided parts of Bengal have deep cultural bonds, on the other, and two divided parts of Punjab have same strong cultural bonds. Those who migrated from Punjab, Bengal, Bihar and east U.P. from India to Pakistan/Bangladesh and vice versa, they too can not forget their ancestors places and keep longing for it even after many generations … India and Pakistan have remained in

---

2 This choice of material reflects our argument that membership on any site is not indicative of wider representation because a moderator may control who joins the group and thereby indirectly control the comments posted. The moderator may also directly delete comments. Open forums with little or no moderation are thus representative of a wider demographic.

perpetual state of anxiety and tensions throughout because of various factors, mostly at ruler's level in both countries.

This forum asserts the voices of common people in India, Pakistan, and Bangladesh, people who want peace and friendship between the three nations, which have deep common cultural bonds and whose citizens have kinship relations. Premchand, Faiz Ahmad Faiz, and Kazi Nazrul Islam – the greatest writers of the region in Hindi, Urdu, and Bengali, respectively – are identified as true representatives of the common people in the region, symbolizing their social life and shared cultural heritage. The forum also urges people not to be indifferent to what is happening with the three national governments. Rather, participants are urged to speak for themselves. This new rhetoric argues that the destiny of the three nations cannot be left in the hands of a few politicians, bureaucrats, and the military in power. In fact, the forum claims that it is the birthright of people of nations to shape their own destiny. Clearly, the forum attempts to produce a narrative of the nations that is written by the people living within them, that hinges on a cultural system or "a system of cultural signification" (Bhabha 1), and that reflects the social life of the people rather than the political ideologies of a few people in power (cf. Anderson).

The Facebook group India Pakistan Friendship Club[3] equally challenges the dominant national master narrative. In its introduction, the club asks,

> Why shouldn't we be friends? Lahore is nearer to Delhi than Washington or London. We look alike, we have similar cultures, and we speak the same language. We have been united in our struggle for freedom for more than 300 years. But somewhere we got misguided, misled and we became prey to the machinations of an outsider.
> If you believe in this, then this forum is for you.
> Lets be part of the solution ...
> Peace!

The notion of being misguided and misled in this quotation refers to the dominant master narrative, which is underlined by a deeply masculinist view that stereotypes the two nations, India and Pakistan, as binary opposites deeply fragmented by religion and ideology. Any differing narrative is immediately stigmatised and labelled as "soft," meek, and

---

3 India Pakistan Friendship Club is a Facebook group with over 2,000 members; https://www.facebook.com/groups/IPFC.FB (last accessed October 2014).

anti-nationalistic. Thus, it is not difficult to understand why the dominant master narrative not only controls public imaginations but also emerges as the "official" state-backed narrative despite the discourse of secular liberalism espoused by the founding fathers, Jinnah and Gandhi, of the two nations (Azad). The dominant master narrative here suppresses and rejects different forms of knowing; it sees and reimagines the nations as soft, characterizing itself as masculine and alternative narratives as feminist. It is these different forms of seeing and reimagining the world that Cornell argues are the core of the feminist project. By contrast, the almost subterranean counter-narratives in quick media, particularly social networking sites, stem from personal experiences or deeply personalized encounters and family contexts.[4] These counter-narratives are not views from nowhere, nor are they impersonal, abstract, rationalist accounts or responses in the public realm. In fact, they have personal contexts and are situated in specific histories, identities, emotions, and attachments. In other words, they are views from somewhere (Wittknower).

Our feminist lens for understanding transnational identities is rooted in the notion that meanings and constructions of identity are located in deeply personal contexts but are marginalized for their so-called subjective constructions and hence must be relegated to the realm of the private. Feminist critics have long argued that the binary separation of the public and the private has been sequestered along the lines of gender (Rosaldo and Lamphere; Reiter). The very presence of these subjective feminist constructions in the public is more often than not thought of as a challenge to dominant nationalistic master narratives, one that threatens the lines of separation – between the nation-states on one hand and between the public and private on the other.[5] More so, the modern

---

4 The argument is drawn from Elisabeth Porter and Carol Gilligan, who argue that the contribution of feminism to ethics is the emphasis given to personal experience, nurturing relationships, emotional judgments, and particular contexts.

5 In Western political and philosophical traditions, the public/private dichotomy is perhaps more pronounced and debated. The public/private dichotomy is often associated with notions of masculinity and femininity. The public embodies rationality, logical thinking, and objectivity – qualities that are identified as being inherently male. On the contrary, the private is emotional and subjective, which are supposedly intrinsically female in nature. Thus, naturalizing the public and private arenas in such a manner that the public then becomes "naturally" the domain of the men and the private that of the women causes the presence of women in public to be regarded as "out of place." See Rose, Massey for discussions on space being another arena where gender is constructed and performed.

nation-states have undergone a process of feminization where women are central to the act of visualizing and conceptualizing the nation (Ray). Women have become the "national iconic signifier[s]" who are "worshipped, protected, and controlled" (Kaplan, Alarcón, and Moallem 10; Sarkar). Thus women are to be encoded within the borders of the nation-state and consigned to the private (Sangari and Vaid; Spivak; Shohat; Kaplan, Alarcón, and Moallem).

Not surprisingly, women and quick media emerge as subversive sites through which the meanings, values, and borders of the nation-states are produced and reproduced, contested and simultaneously reinforced (Berry). The following sections show how both women and quick media define and redefine borders and identities of nation-states in South Asia. We do so with a close analysis of the public debates around tennis star Sania Mirza in quick media. In South Asia as elsewhere, identities, whether national or otherwise, are rarely unified. Identities are multiple and ambiguous, and the presence of quick media produces a dialogic space through which the meanings of these identities are negotiated and new connections made in the context of shared histories and cultures. The negotiations and new connections rework and articulate notions of family and kinship that challenge traditional concepts of family and suggest the possibility of choosing one's family based on newer criteria. Thus emerges the idea of an imagined family based on one's own choosing, disrupting the traditional model. The example of Sania Mirza presents insight into the blurring of national, religious, and familial boundaries.

Sania-tizing the Nations: Of Blurring B/orders and Identities in South Asia

> Every word I speak, every skirt I wear is discussed and analyzed … Wherever I go, people look at me. That is why these days I prefer to stay at home. I have to learn to live with all this. It is quite disturbing that my dress has become the subject of controversy; I don't want to say anything on this. (Sania Mirza, quoted in Mir, n.pag.)

Sania Mirza is an Indian Muslim tennis player who has won several grand slam tournaments around the world and is considered a celebrity off the tennis court; Wikipedia named her one of the top 50 Asian heroes of 2005. But despite – or perhaps because of – her achievements, Mirza has persistently been at the centre of controversies.

Mirza's narrative of being "discussed," "analysed," and "looked at" and her choice, as a result, to "stay at home" presents an example of how women are essentially controlled and confined to the private domain. However, Mirza's positioning as a celebrity shows her identity and her body as the nation-state's iconic cultural and territorial signifier.

What interests people most about Mirza is her tennis career, her sense of fashion, and her marriage to Shoaib Malik, a well-known Pakistani cricket player. Comments about Mirza's marriage to Malik show the degree to which personal relationships and national narratives are interwoven in the realm of quick media. While traditional media focused on a narrow view of Mirza, emphasizing coherent identity markers, in contrast, quick media opened up a dialogic space that made possible contact and conversations across the borders, highlighting both agreement and conflict.

Mirza presents an interesting case study as she in many ways challenges b/orders – as a woman, as a Muslim, as an Indian, and as the wife of a Pakistani. Controversies arise because she often transgresses boundaries with ease but epitomizes the blurring of these b/orders. Mirza's performance questions the locational politics articulated through the ordering and othering processes that borders tend to signify and contain. Notably, Mirza enjoys great popularity in both India and Pakistan despite the controversies and particularly in quick media, where everyday conversations about her take place across these borders ("Controversy Queen"). Conversations in quick media focus on Mirza and her personal life, especially her transgression of borders, thus facilitating the blurring of borders and formation of kinship ties.

The Female Body

In September 2005, Mirza encountered the first controversy – that of being a Muslim woman and yet wearing a miniskirt while playing tennis. Her attire was seen as provocative, exposing her body in public and to the voyeuristic and scopophilic male gaze (Hubbard; Mulvey). Specifically, some argued that Mirza's attire was anti-Islamic.[6] The ensuing furore in quick media within the Muslim community in both India and Pakistan

---

6 The clergy of Jamiat-Ulema-e-Hind (Islamic organization in India) registered their objection against the dress code of the tennis player, arguing that it is against Islamic faith.

first brought to light the complex and paradoxical positions that result from the blurring of b/orders drawn by Indian and Pakistani masculinist nationalist discourses. Some of the blogs reflect on the regulation of Mirza's femininity. For instance, commenter Sobia writes,

> What irritates me about all the commenting on Sania Mirza's clothing is this assumption that somehow these men have a right to her body. That these men think that they have the right to tell her how to dress demonstrates that they think they have the right to control her body. And that disgusts me. Her body is her body and she can do with it what she chooses. Not only that, but her religiosity is only her business. No one else has the right to tell her how to express her religiosity. (comment on "Pride or Disgrace")

Sobia's narrative comes from the perspective of a woman of Pakistani origin and draws out her resistance to the nationalist narrative that tries to control and reduce a woman's body. Sobia specifically looks at Mirza's body as being the discursive site where boundaries of private versus public and of national religious identities are reproduced (Ramaswamy). The voicing of resistance by Sobia, a Pakistani national, on behalf of Mirza, an Indian Muslim, reconfirms the connectedness between the nations and their people. Likewise, Indscribe voices his/her resistance to the stereotypical production of a conservative, regressive Islam. Indscribe rejects the ways that Indian Muslims are being painted as a result of Islamic groups' opposition to Mirza's attire, emphasizing that a "couple of people" are not representative of the entire Muslim community. Indscribe writes, "What a joke. Sania Mirza was criticised just by a couple of people (among 200 million Muslims) and they were also condemned by Muslims. She is a role model and hero. Indian Muslims are not at all such bigots as you are trying to make them up to be" (comment on "Pride or Disgrace"). Both narratives highlight shared Islamic values that seem to tie the two nations together in a bond of kinship and an understanding of the countries' heterogeneity. As such, both narratives destabilize the notions of a conservative fundamentalist Islam through their resistance to the manner in which religious ideologies are being used in disciplining women. In fact, the two quotations also show how Muslims across the borders of India and Pakistan speak a similar language of resistance produced from identical experiences and contexts. These shared experiences within the contemporary context and their similar languages of resistance open up a common ground for further conversations to take place and develop an understanding of each other's contexts. A kind of bond and relatedness, both familial and kinship, which unites the people of the two nations materializes, born from knowing and understanding each other's contexts, experiences, and struggles.

Sexuality

Shortly following the furore regarding her apparel, in November 2005, Sania Mirza stirred controversy by voicing her support for South Indian film actress Khushboo and her campaign for safe sex. Mirza's support was seen in mainstream media as promoting open sex and premarital sex and defying Indian cultures, values, and ethics. Interestingly, however, Mirza received significant support online, particularly across the border in Pakistan. The following conversation on a Pakistani Web-based discussion group reveals the deep-rooted understanding that people possess as a result of their shared contexts and experiences and their similar history and culture. This forms and evokes a feeling of relatedness, a prior connection, and a sense of belonging and of having kinship ties that are beyond blood relationships.[7] These conversations attempt to contextualize Mirza's controversial comments within a contemporary society that is faced with real threats of sexually transmitted diseases from unsafe sex. On the other hand, the commenter recognizes Mirza's remarks as an individual's personal opinion and not as an "official representative of Islam." Faisal writes,

> This is what she said (according to *New Kerala Times*): "Look, whether it's before or after marriage, people should have safe sex. And about pre-marriage sex, you can't stop people and hence the best way is to play it safe." The topic is safe sex, to prevent AIDS/HIV and other STDs. There are people in all cultures who indulge in pre-marital, extra-marital and post-marital sex. If everyone only has sex with their spouse then there won't be such rapid spread of STD's etc. The way I look at it, she is making a point for all those people who engage in unsafe sex to use protection. To turn it around and suggest she is promoting pre-marital sex is a classic disinformation technique.

---

7 Lyon (2002) argues that Pakistan is a unique cultural area because of its hybridizing society having Indian influences: linguistically, in terms of the usage of Urdu and Hindi; and socially, in terms of the syncretic Islam practised in Pakistan bearing similarities to Hinduism. See also Saideman and Ayres (2008).

To this, PakistaniAbroad adds,

[W]hat's up with the mullahs ... they are so damn obsessed by the next skirt that comes along that they have to hang on to everything she says ... who the eff cares what she thinks or says? is she some kind of Mullani? where did she say she's going to be the official representative of Islam? why should anyone care what she thinks?

The dynamics of the conversation can also be seen in the following chat log on Gupshup between Lahore981 and Faisal over the course of a few hours:

LAHORE 981: What I don't understand is that these people who enjoy giving such statements about Islam and Islamic provisions are too self centered and too arrogant to hit Islam. Probably they in themselves are quite proud, and have forgotten Allah to say such things in public. What does she think of herself? She has become a tennis star but has she become a goddess? Has she grown golden horns on her head by becoming a tennis star? Such are the people who hold a token for extreme punishment in the hereafter ... and it doesn't matter if they call themselves Muslims or have Muslim names.

LAHORE 981: Aha. Whatever the explanations [that] have been given in this thread, it [would be] better for her to phrase her intent in a little better words. Her words are too controversial to believe her intent was to convey the message "Safe sex" Huh ...

FAISAL: This is what she said: "Look, whether it's before or after marriage, people should have safe sex. And about pre-marriage sex, you can't stop people and hence the best way is to play it safe." When asked to talk about safe sex and spread of AIDS/HIV, how would you have said it Lahore 981? Just remember you are not talking about (or to) a Muslim audience, but you are going to comment on a largely non-Muslim population of India.

Mirza's comments on safe sex have been criticized by Muslims as anti-Islamic, as seen in Lahore981's remarks that again refer to a set of common Muslim/Islamic values. Yet Faisal defends Mirza's comments, stating that they are not just meant for a Muslim audience but need to be seen in the context of addressing a largely non-Muslim Indian population. The recognition of a need to view a Muslim in a non-Muslim context differently and the suggestion that Mirza become aware of the need to reword her comments despite her good intentions erase the totalizing boundaries that the dominant master narrative tends to create. Simultaneously, the narratives talk of an insight into the Indian context which reflects a certain connectedness between the two nations that forms the basis of forging kinship ties. In fact, it accepts and gives recognition to the creolization of cultures, in this case a Muslim living in non-Muslim, Hindu contexts.

Patriotism

Almost all events around Mirza became markers that drew and redrew the borders of the nation-states on one hand and questioned her nationalistic sentiments, her loyalties, and her religious identities on the other. This blurring of borders created an in-between dialogic space for the conversations in quick media not only to take place but to form a kind of community and kin that transcended the boundary of the nation-state. Mirza's decision to opt out of her partnership with Israeli player Shahar Pe'er in 2006 for tennis tournaments in anticipation of protests that might be made by Muslims across the Muslim world, and particularly in India and Pakistan, highlights that she created a connection that fostered kinship ties among Muslims in India and in Pakistan. In both India and Pakistan, many Muslims, and radical Muslims in particular, consider Israel an anti-Islamic country. Sania Mirza's actions were criticized heavily through comments such as this: "Shameful! Really shameful that a sportsperson succumbs to a bunch of losers who would do anything to get their way (in the name of their faith). Why is she following something she does not believe in? [Does] everything she does on and off [the] court comply to the whims and fancies of these jokers?"[8] These critiques mainly come from non-Muslim Indians who thought that by dropping out of her partnership she compromised her chances of winning and, in doing so, compromised the chances of India's victory. This incident underlines how Mirza's actions define nationalistic sentiments and how they seemingly contest with her Muslim identity not just in India but in the larger Muslim world.

---

8 See the comments by the readers on a news article regarding Mirza's decision not to team up with an Israeli tennis player at http://www.ynetnews.com/articles/ 0,7340,L-3217231,00.html (22 Feb 2006).

In December 2007 Mirza was accused yet again of being "un-Islamic" and "hurting" the sentiments of the Muslim community for shooting an advertisement in the Mecca mosque in Hyderabad. Similar accusations of being "unpatriotic" and "hurting" Indians were thrown at her when she was pictured watching her compatriot Rohan Bopanna play in the Hopman Cup (in 2008 in India) in a resting pose with her feet on the table just next to the Indian national flag, or when she refused to play in local tennis tournaments in India, namely the Bangalore Open, in the same year. She even faced possible prosecution under the "Prevention of Insults to National Honour Act." The production of the dominant master narrative created parallel dissenting voices in quick media, and the ensuing narratives further produced more complex and fuzzier b/orders of nationalistic identities and opposition to the binary constructions of India as a Hindu nation and of Pakistan as a Muslim one. The following comments posted on online articles about Mirza illustrate this dynamic.

In response to "Sania Mirza's Showing Disrespect to Indian Flag Questionable," Bobby writes,

> I am sure that only empty headed fools like the advocate who filed the case against Sania Mirza believe that Sania can insult India. In fact Sania is one of the greatest patriots in India. She is bringing soo much fame and pride to India, that some people are not able to digest that a girl can do so much for India. It is a shame that in our country useless people are able to misuse the court to harass productive people, while molesters and rapists are let free by our system. I wonder why some people are trying to bring down achievers? Maybe they don't like people who bring fame to India. I believe that sane people will understand who are the real antinationals. I wonder what people mean by "India shining,"[9] while in reality issues like this makes India blush with shame in front of the world. (Vijay, n.pag.)

On the same article, Leo states,

> Those people who are filing cases against Sania should think what good things they are doing as a good Indian citizen. Instead of wasting their

---

9 "India Shining" was a marketing slogan referring to the overall feeling of economic optimism in India in 2004. The slogan was popularized by the then-ruling Bharatiya Janata Party (BJP) for the 2004 general elections.

time in filing such kind of rubbish cases and demoralising people, they should be doing some good if in reality they are so worried about national pride. There are so many people dying out of hunger, so many children not getting education. So [please] think about all these problems instead of taking advantage of the big name of celebrities.

On an article titled "Sania Had Thought about Retiring," an anonymous reader posted,

I feel Sania is more disturbed over the charges of disrespect to Tricolour than skirt or Mecca masjid issue. She really felt hurt [because] some miscreants are raising a question on her patriotism which is totally unacceptable. The name and fame she has brought for India is a remarkable story and it does not need anybody's certificate. There is a famous saying in English that "Patriotism is the last resort of a scoundrel" so nowadays unfortunately every crook is trying to prove that he/she is the biggest patriot and others are traitors. Absolutely funny. It is an irony for India that those who are involved in mass murders, rapes riots and disturbing the social fabric of India, have become heroes of nationalism whereas those who are honest, simple and believe in harmony have became an object of suspicion. Great.

Reader Satwant Kaur voiced similar concerns in a comment on "The Sania Row":

Sania's decision, presumably because of the tension arising from the controversy over her dress and alleged disrespect to the national flag, only reflects society's attitude towards women. Why are our religious feelings so easily hurt by the dress, art works and writings of women and why do we feel so outraged by presumed acts of our icons showing disrespect to our national symbols? Why aren't we worried about more serious issues such as corruption, poverty, unemployment and lack of proper educational and health facilities?

Clearly, these narratives highlight the tensions within the dominant master narrative. These comments question the basis for criticizing Sania Mirza, who has, to most of the commentators, established her patriotism and love for the country by representing India in the international tennis circuit and making India proud. The narratives question what defines patriotism and the claim towards an Indian identity.

None of the narratives highlight Mirza's religion, but the very silence regarding her religious identity indicates a counter-narrative that moves away from considering religion as the defining criteria for either patriotism or national identity. The silencing of Mirza's religious identity underlines the ambivalences in the making of national identities that perhaps crafts possibilities for forming newer kinship ties.

Personal Life

Mirza's personal life became a site of subversion of the dominant master narrative of the divided nations. In early 2010 she broke off her engagement to childhood friend Mohammad Shorab Mirza, an Indian, to later announce her wedding to Pakistani cricket player Shoaib Malik. The news about her relationship went viral in both mainstream media and quick media, particularly after the news that Shoaib Malik was already engaged or perhaps even married to another woman from Hyderabad, Ayesha Siddiqui. Mirza's marriage was mired with controversies, with allegations and counter allegations coming at her from all quarters, legal notices being served, and community leaders, elders, and politicians being involved in resolving the issue informally. The marriage became a politically contentious issue with two right-wing political parties, BJP (Bharatiya Janata Party) and Shiv Sena, criticizing her decision to marry a Pakistani as unpatriotic.[10] Mirza's marriage was interpreted as an act of betrayal against the nation itself because it destabilized the political unity and the imagined oneness of India by entering into an intimate relationship with the "other," suggesting Pakistan as a hostile nation, in keeping with the master narrative.

Mirza's marriage opens up the border facilitating the building of lasting personal and intimate relationships between the two nations. Her marriage, many claimed in social media, "[s]hows, however [big] the political differences and mental differences between Pakistan and India somehow [they are] still bound by the same culture, same blood and love for each other."[11] The notion of "same culture" translates into ideas about "same blood" and "love for each other," which evoke feelings of

---

10   https://www.facebook.com/pages/shoaib-malik-sania-mirza/106679019366442 (5 April 2010, 7:57 a.m.).
11   "Sania Mirza to wed Shoaib Malik?" at https://www.facebook.com/pages/Sania-Mirza-to-wed-Shoaib-Malik/108663969162061.

kinship and familial bonds. Others in social media are equally quick to point out that kinship and family ties are not just limited to marital relationships. Families have roots on both sides of the border predating the partition of India in 1947. Both Indians and Pakistanis before partition had ancestral homes and recollect stories and memories of these places on either side of the border (Rothberg). These remembrances create a sense of belonging. Shared memories tie in relationships of kinship that question the relevance of national borders (Hall). For example, one commenter writes, "So why not remove the facade of enmity or be obsessed with limited goals. Instead, we can aspire for better life for the people on both sides and make the borders (both international borders and Line of Control) irrelevant (still remaining as sovereign countries). Let the culture and common history win."

Despite some controversy, people across the borders rejoiced and congratulated the married couple, especially in social media, as if they were a part of a large family. Likewise, jokes were made and people teased the newlywed couple with statements like "Dil Walay Dulanhiya lay jain gay" (Large-hearted men always win the bride) and "Bagha kai tori laa rahay hai shaadi kar kai laa rahay hai" (Don't elope but marry and bring her home). Many Pakistanis also started referring to Mirza as *Bhabi* (sister-in-law),[12] even before marriage. People even advised the newlywed couple on the intricacies of marital relationships as is done traditionally in families: "All married couples should learn the art of battle as they should learn the art of making love. Good battle is objective and honest – never vicious or cruel. Good battle is healthy and constructive, and brings to a marriage the principle of equal partnership."[13]

The Sania-Shoaib marriage and its reception bring a new perspective to the relationship between India and Pakistan, tying the two nations in a much more intimate familial bond. It also re-examines the dominant master narrative through new interpretative lenses as exemplified in the previous narrative. The basis of any strong relationship is equal partnership, which is based not just on love but also the ability to fight a good battle. The dominant master narrative is perhaps the "art of battle" while the counter-narrative in quick media is the "art of making love," and both complement each other to create a marriage of equal partnership between India and Pakistan.

---

12 https://www.facebook.com/saniamirzafanclub (4 April 2010, 3:00 a.m.).
13 https://www.facebook.com/pages/SANIA-MALIK-Queen-of-Pakistan/
    115935165090057 (14 April 2010, 5:12 p.m.).

Conclusion

As this case study of Sania Mirza shows, a peculiar anomaly emerges: on one hand is the widely held belief and public discourse about India and Pakistan as divided nations with differing religion, ideologies, cultures, and identities. In mainstream contexts any counter-discourse does not have any space and is almost subordinated by the hegemonic master narrative. This dominant narrative construction is suggestive of the manner in which hegemonic systems of masculinist power are maintained and naturalized and how they contain and sublimate the other, the counter-narratives in this case. In Bhabha's view, the counter-narratives highlight the irredeemable plural character and the ambiguities in the essentialist and totalizing nature of the nation and its imagined community, which the hegemonic master narrative continually tries to evoke and create.

Yet this case study suggests that there is a tacit recognition of the space that the counter-narratives have created in and through quick media, subverting these very hegemonic masculinist narratives: for example, mainstream discussions of Bollywood and cricket as the *new* religion in South Asia, and in India and Pakistan in particular, or the fact that in a recent issue of *Outlook*, a mainstream Indian magazine, a Pakistani historian writes (or rather rewrites) the history of an undivided subcontinent with Jinnah as its first prime minister ("The Jinnah Tapes"). The question that remains unanswered is whether the counter-narrative is indicative of a sudden change in the relationship between the two nations and in how the two nation-states and their identities are constructed. If such a change is occurring, what is it about the present moment that has allowed this shift to begin? We argue that the counter-narrative was perhaps always present but has invented and reinvented itself in many forms, all of which play a pivotal role in the narrative of the making of the two nations. This reinvention has been facilitated through new media paving the way for the "network society" (Castells) of sharing spatiality, albeit virtually. It is this porousness that produces the "postnational" (Appadurai) that helps to reconstruct and bring to the fore many histories, memories, and shared experiences and relationships of old. It is these many histories, memories, experiences, and cultures that are shared by the people of India and Pakistan through quick media which form the basis of relatedness and newer definitions of kinship. Quick media, we argue, open up a dialogic space to share and forge kinship ties that tend to go beyond the latent biologism of kinship, making it a sociocultural phenomenon (Carsten; Franklin and McKinnon; MacCormack and Strathern; Collier and Yanagisako; Schneider). Similarly, the dialogic space of quick media has produced newer narratives of solidarity and coexistence between India and Pakistan.

Therefore, we conclude that the multiple histories and shared experiences, memories, and cultures forge new kinship ties, which in turn create new ways for people to relook and relocate in the larger context of South Asia. Quick media give birth to spaces of alterity, in terms not only of producing counter-narratives that are subversive, but of creating continuities with forgotten old relationships, forging new ties of kinship that have been either diluted or marginalized by the overarching masculinist official nationalist narratives.

WORKS CITED

"107th Birthday of Bhagat Singh Celebrated." *Daily Times of Pakistan* 29 Sept 2013. Web. 4 Mar 2014.

Akbar, M.J. "India and Pakistan: The Great Wall of Silence." *India Today* 10 Aug 2012. Web. 4 Mar 2014. <http://indiatoday.intoday.in/story/india-today-editorial-director-m.j.-akbar-india-pakistan-65-years-of-independence/1/212768.html>

Anderson, Benedict. *Imagined Communities: Reflections on the Origin and Spread of Nationalism*. London: Verso, 1983. Print.

Appadurai, Arjun. *Modernity at Large: Cultural Dimensions of Globalization*. New Delhi: Oxford University Press, 1996. Print.

Azad, Maulana Abul Kalam. *India Wins Freedom*. New Delhi: Orient Blackswan, 2009. Print.

Berry, David M., ed. *Understanding Digital Humanities*. New York: Palgrave Macmillan, 2012. Print.

Bhabha, Homi K., ed. *Nation and Narration*. London: Routledge, 1990. Print.

Carsten, Janet. *Cultures of Relatedness: New Approaches to the Study of Kinship*. Cambridge: Cambridge University Press, 2000. Print.

Castells, Manuel. *The Rise of Network Society*. Cambridge: Blackwell Publishers, 1996. Print.

Collier, Jane Fishbourne, and Slyvia Junko Yanagisako, eds. *Gender and Kinship: Essays Towards a Unified Analysis*. Stanford: Stanford University Press, 1987. Print.

Cornell, Drucilla. "What Is Ethical Feminism?" *Feminist Contentions: A Philosophical Exchange*. Ed. Seyla Benhabib, Judith Butler, Drucilla Cornel, and Nancy Fraser. New York: Routledge, 1995. 75–105. Print.

During, Simon. "Literature – Nationalism's Other? The Case for Revision." *Nation and Narration*. Ed. Homi K. Bhabha. London: Routledge, 1990. 138–53. Print.

Franklin, Sarah B., and Susan McKinnon. *Relative Values: Reconfiguring Kinship Studies*. Durham, NC: Duke University Press, 2001. Print.

Gilligan, Carol. *In a Different Voice: Psychological Theory and Women's Development*. Cambridge, MA: Harvard University Press, 1983. Print.

Gupshup. Forum. n.d. Web. 4 Mar 2014. < http://www.paklinks.com/ gs/religion-and-philosophy/199460-little-sania-favors-pre-marital-sex. html>

Hall, Stuart. "On Post-modernism and Articulation: An Interview with Stuart Hall." *Journal of Communication Inquiry* 10.2 (1986): 45–60. Print.

Hubbard, Phil. "Sex Zones: Intimacy, Citizenship and Public Sphere." *Sexualities* 4.1 (2001): 51–71. Print.

"The Jinnah Tapes 1." *Outlookindia.com* 9 Sept. 2013. Web. 4 Mar 2014.

Kaplan, Caren, Norma Alarcón, and Minoo Moallem, eds. *Between Woman and Nation: Nationalisms, Transnational Feminisms, and the State*. Durham, NC: Duke University Press, 1999. Print.

Lankala, S. "Mediated Nationalisms and Islamic Terror: The Articulation of Religious and Postcolonial Secular Nationalisms in India." *Westminster Papers in Communication and Culture* 3.2 (2006): 86–102. Print.

Lyon, Stephen M. "Local Arbitration and Conflict Deferment in Punjab, Pakistan." *Anthropologie* 40.1 (2002): 59–71. Print.

MacCormack, Carol P., and Marilyn Strathern, eds. *Nature, Culture and Gender*. New York: Cambridge University Press, 1980. Print.

Massey, Doreen B. *Space, Place, and Gender*. Minneapolis: University of Minnesota Press, 1994. Print.

Mir, Shabana. "Sania Mirza's Clothes: What's Really Going on Here?" *Altmuslim: Global Perspectives on Muslim Life, Politics and Culture* 29 Sept 2005. Web. 4 Mar 2014.

Mulvey, Laura. "Visual Pleasure and Narrative Cinema." *Screen* 16.3 (1975): 6–18. Print.

Porter, Elisabeth. *Feminist Perspectives on Ethics*. London: Longman, 1999. Print.

Ramaswamy, Sumathi. *The Goddess and the Nation: Mapping Mother India*. Durham, NC: Duke University Press, 2010. Print.

Ray, Sangeeta. *En-gendering India: Women and Nation in Colonial and Post-colonial Narratives*. Durham, NC: Duke University Press, 2000. Print.

Reiter, Rayna R. *Toward an Anthropology of Women*. New York: Monthly Review Press, 1975. Print.

Rosaldo, Michelle Zimbalist, and Louise Lamphere, eds. *Women and Culture and Society*. Stanford: Stanford University Press, 1974. Print.

Rose, Gillian. *Feminism and Geography: The Limits of Geographical Knowledge*. Minneapolis: University of Minnesota Press, 1993. Print.

Rothberg, Michael. "Locating Transnational Memory." *European Review* 22.4 (2014): 652–6. Print.

Saideman, Stephen M., and R. William Ayres. *For Kin or Country: Xenophobia, Nationalism and War*. New York: Columbia University Press, 2008. Print.

Sangari, Kumkum, and Sudesh Vaid. *Recasting Women: Essays in Indian Colonial History*. Delhi: Kali for Women, 1999. Print.

"Sania Had Thought about Retiring." *Times of India* 4 Feb 2008. Web. 4 Mar 2014. <http://timesofindia.indiatimes.com/opinions/2756289.cms?comment type=mostdiscussed>

"Sania Mirza: A Pride or Disgrace to Indian Muslims." *Muslimah Media Watch* 10 Jan 2013. Web. 4 Mar 2014. <http://www.patheos.com/blogs/mmw/2013/01/sania-mirza-a-pride-or-disgrace-to-indian-muslims/>

"Sania Mirza – Controversy Queen." *Zeenews India* 31 Mar 2010. Web. 4 Mar 2014.

"The Sania Row." *The Hindu* 8 Feb 2008. Web. 4 Mar 2014. <http://www.thehindu.com/todays-paper/tp-opinion/the-sania-row/article1195082.ece>

Sarkar, Tanika. *Hindu Wife, Hindu Nation: Community, Religion and Cultural Nationalism*. London: Hurst, 2001. Print.

Schneider, David Murray. *A Critique of the Study of Kinship*. Ann Arbor: University of Michigan Press, 1984. Print.

Shohat, Ella. "Territories of the National Imagination." *Transition* 53 (1991): 124–32. Print.

Spivak, Gayatri Chakravorty. *In Other Worlds: Essays in Cultural Politics*. London: Methuen, 1987. Print.

Varshney, Ashutosh. "India, Pakistan and Kashmir: Antinomies of Nationalism." *Asian Survey* 31.11 (1991): 997–1019. Print.

Vijay. "Sania Mirza's Showing Disrespect to Indian Flag Questionable." Merinews.com 13 Jan 2008. Web. 4 Mar 2014. <http://www.merinews.com/article/sania-mirzas-showing-disrespect-to-indian-flag-questionable/129280.shtml#sthash.POd5LdYv.ERPyKLha.dpuf>

"War Chhod Na Yaar: Faraz Haider Hits the Jackpot in His Debut Directorial." *Times of India* 4 Oct 2013. Web. 4 Mar 2014.

Wittknower, D.E., ed. *Facebook and Philosophy: What's on Your Mind?* Chicago: Open Court, 2010. Print.

Yousaf, Nasim. "Pakistan and India: The Case of Unification." Paper presented at the New York Conference on Asian Studies, Cornell University, 9–10 Oct 2009.

Shaping Identities

4 Queering "Web" Families: Cultural Kinship through Lesbian Web Series

JULIA OBERMAYR

In times of seemingly equal rights for women in the Western world, it might be hard to accept that even today, narratives about complex female characters are still underrepresented in the majority of the film industry. We, as a diverse but equally conditioned audience, are repeatedly trained to overlook this fact by being introduced to on-screen women who, at a closer look, turn out to be created mostly in relation to main male characters. Since the introduction of the Bechdel Test by Alison Bechdel and her friend Liz Wallace through the comic *Dykes to Watch Out For*, viewers can check this for themselves by watching any kind of movie, TV show, or Web series[1] with a certain awareness (cf. Sarkeesian, "The Bechdel Test"). The Bechdel Test exhorts viewers to look at movies from a different point of view, which assumes that most productions worldwide are dominated by a certain male gaze, even when produced by women. In her comic strip, Bechdel captures Wallace's idea of analysing how long at least two named women (or girls) have

---

1 "Web series" (sometimes also called Web shows) is mainly used by the online community, journalists, and filmmakers and therefore kept as a term in this article. The term "series," however, is not always correct when applied to Web series. A "series" is usually divided into episodes, telling a different self-contained story, while a "serial" has a plot that continues through the episodes, gradually revealing more details (cf. Cardini 115). Web series tend to mix forms and are of short duration. Furthermore, Web series, like regular TV series, consist of seasons that are structured in episodes. Additionally, the genre varies as well. While drama is the most common, crime, comedy, and the relatively new genre of reality shows are all subsumed under the term "Lesbian Web Series."

a conversation (or a scene in general) without mentioning or focusing on men. Interestingly, a great majority of blockbusters, Oscar winners, and other well-known movies do not pass the test, as Anita Sarkeesian, media critic and creator of the YouTube channel Feminist Frequency, contends ("The Oscars"). She further claims that the test does not give any evidence of quality but simply highlights the construed presence of women as only existing in relation to men. Because of numerous debates about very brief interactions between female characters in films which could pass the Bechdel Test, Sarkeesian feels the need to add another rule to the test: these interactions have to be at least 60 seconds long ("The Oscars"). Sarkeesian further remarks that films at the Oscars mostly pass if they are already female centred.

This chapter seeks to explore a specific kind of female-centred media production that easily passes the Bechdel Test[2]: Lesbian Web Series, that is, a serial of scripted lesbian-focused shows which are primarily released on the Internet as opposed to through television. As they negotiate a diverse range of non-normative identities as well as reinforce a strong audience engagement and interaction through quick media, Lesbian Web Series[3] function as a new community-building and community-altering technology, claiming virtual space for their imagined families. Streaming their series online, Web series creators frequently use video platforms or embed their shows in websites that offer a comment section or guest book for viewers to leave feedback or other thoughts. There is no doubt that quick media influence the production and dissemination of Lesbian Web Series since they mainly operate within social media, rarely investing in TV ads. This chapter tackles the question of what kind of identity affirmation is provoked by

---

2   The Bechdel Test internationally provokes an important shift in media perception: female characters independently get attention. In Sweden, movie theatres even started to implement passing the test as a criterion for films to get the highest ranking "A" (cf. dieStandard.at editors).

3   Lesbian Web Series, as I suggest calling them, since it has already become a common term particularly among the audiences of the same, are not always labelled as such by their creators. The term "lesbian" often indicates a niche product which (unwillingly) targets a smaller group of viewers. The recent trend towards the denomination or addition of "queer" instead of "lesbian" or "LGBT," which can be detected not only in headlines of scholarly papers but also in the renaming of LGBT organizations, does occur in Lesbian Web Series as well, although less frequently given that Lesbian Web Series, according to my definition, need lesbian protagonists, whereas Queer Web Series do not.

the series and how viewers negotiate it in their commentaries and form kinship among other members of the audience.

Lesbian Web Series: A Tool for Identity Reinforcement?

Before jumping into the analysis of identity negotiation in viewer comments, a few things have to be said about Lesbian Web Series and their core functions beyond simple entertainment: they ensure a diverse lesbian visibility with high dissemination/accessibility worldwide, nurture a participatory (sub)culture by using quick media, and offer a virtual safe space for female-centred community building, identity formation, and kinship.

Lesbian Web Series are serial online entertainment of any broadcasting format or genre where the leading part is primarily dedicated to one or more lesbian characters, even though the representation of those characters can differ regarding their lesbianism. Some, for example, mainly concentrate on the struggles and difficulties of the coming out process of their teenage characters (e.g., *Out With Dad*), while others take their protagonists' lesbianism as a given (e.g., *Venice the Series*), targeting other topics besides homosexuality or consciously playing with sexual fluidity (e.g., *Chica Busca Chica*). They strategically employ a particular set of cultural markers that is mainly recognized by gay women, such as specific music with lesbian lyrics or references to movies with lesbian content. Therefore, Lesbian Web Series provide not only visible representations of lesbian identities but also a means of identification for those viewers who are able to anticipate and/or decode these cultural markers.

This chapter specifically investigates how American, Canadian, and Spanish Lesbian Web Series facilitate kinship among the audience despite different cultural backgrounds. While the various series reference different national and cultural content, their depiction of lesbian subcultures connects audiences transnationally through the Web. How YouTube, or more specifically its comment function, contributes to the formation of a first kinship among (lesbian) women online is shown with a closer look at the comments posted beneath the episodes of *Chica Busca Chica*, *Venice the Series*, and *Out With Dad*, which serve as examples for each of the main production countries: Spain, United States, and Canada, respectively. In order to examine the audience's commentaries underneath the video entries of these three Lesbian Web Series, the respective corpus includes comments on YouTube for each series'

episode 1 from season 1. These Web series intentionally put the spot-
light on strong and fully developed complex female protagonists with
depth who equal male heroes on TV. They create dynamic female identities rooted in cultural kinship, and their storylines generally unfold
among a close group of women, gay or straight and everything in
between. Furthermore, production mostly takes place locally.

Chica Busca Chica, for example, is set in Madrid, portraying feminine,
lesbian stereotypes such as the player Nines (the Spanish *femme* version
of *The L Word*'s character Shane), who works as a waitress and bartender
in the bar Chica Busca Chica; the challenged and clingy Mónica; the lesbian newbie Ana from the countryside, who finally gets to explore her
identity in the big city; and a possibly bi-curious but straight Carmen
who initially has a boyfriend with whom she plans on living.

In contrast, *Out With Dad* opts for a typical coming out storyline (in
the area of Toronto), probably because the lesbian protagonist, Rose,
is still a teenager. The series shows both possible sides of this very
delicate process concerning identity development: the understanding
and caring father Nathan, but also the very conservative mother of
Rose's love interest, who completely rejects the girl's sexual orientation. Nathan is portrayed as a multidimensional character, trying to
come to terms with his daughter's struggle but at the same time feeling
inadequate at approaching her about this subject in the beginning of
the show.

Venice the Series continues to shape a virtual world of socially highly
respected workplaces for (lesbian) women in Venice Beach, following *The L Word* model of job depiction, meaning lesbian characters in
mostly well-paid and prestigious positions. The protagonist Gina, an
internationally well-known interior designer, runs her own company.
Her assistant Michele evolves from her busy and hard-working right
hand to an interior designer herself. Lara writes books that make it to
the *New York Times* Best Seller list. Ani is a respected photographer, and
Gina's aunt Jane/Guya embodies a gifted spiritual guide with remarkable powers. The male characters are represented in a more diverse
work range, varying from male secretaries and bartenders to rocket scientists and business executives. It is striking that jobs which require a
great amount of creativity recur for protagonists in Lesbian Web Series.
Photography and visual arts especially keep resurfacing either as actual
jobs (in addition to Ani in *Venice the Series*, Carmen's boyfriend Jorge in
Chica Busca Chica is a photographer as well) or at least through paintings or photos in the background.

In contrast to traditional television offerings, Lesbian Web Series are generally exclusively watched online, unless the filmmakers enter into a definitive agreement with a television network to additionally broadcast on TV later on. For example, *Out With Dad* "was created specifically for the web [t]hough it has been broadcast on Rogers TV in Toronto, and will be again throughout the year" (OutWithDad). As there are no rules or regulations in place yet, it solely depends on the person or organization responsible for the Lesbian Web Series' production as to whether there is a subscription fee to watch the series or if they are simply uploaded on a free video exchange website such as YouTube, where they can be watched at any time. Thus, Web series offer filmmakers more freedom in experimenting with language, duration, and structure, but also freedom to explore content that on TV would only be broadcasted with certain restrictions (especially when it comes to the portrayal of same-sex relationships during daytime). The duration of Web series generally varies greatly since there is no predefined structure to which producers have to adhere. Depending on production costs, funding, and time constraints in recording, the common duration of a Web series episode ranges from 3 to 10 minutes (often within the first season), with a possible development into 15 to 30 minutes' length (possibly because of the average attention span on Internet portals online and the former 10-minute length restriction on YouTube). Financial parameters mean that production mostly takes place on location and rarely in studios,[4] and the number of different locations is usually kept to a minimum (Kuhn 56–7). Viewers often stream the episodes without interruption, thus the longest duration of Web series often equals the length of regular TV shows broadcasted in the United States or Canada without commercial breaks (which equals about 30–45 minutes).

It is crucial to understand how the Web series' different production models play a key role in involving the audience in their processes. There is a general tendency for TV shows to migrate to the Web, especially in the North American soap opera genre. *Guiding Light*, for example, ended in 2009 after more than 70 years of uninterrupted

---

4 Due to the relatively new form of online streaming and distributing Web series to a global audience on the Internet rather than a national one on TV, advertisers still remain sceptical about financially supporting Web series. Great numbers of viewers are harder to attract and maintain than on TV and thus not as economically interesting for advertisers (yet). Therefore, cheaper and faster ways of production are necessary to compensate for the smaller budget.

broadcasting (1937–1952 on radio; 1952–2009 on television); since then, numerous other soaps also ended, primarily because of financial issues. Therefore, the entertainment industry as a whole is searching for new ways and means to tell their stories. The Web could provide a solution to budgetary cutbacks, since production costs can be kept very low compared to TV shows. The lack of a larger budget often leads Web series creators to different methods in assembling necessary materials and gathering the people with whom they are working. Some fans, if not already integrated in the work on set, help out with subtitle translations or promoting the show. As translations also happen for illegal digital copies of TV series on the Internet, it is worth mentioning that in the case of Lesbian Web Series, fan "translators"[5] often officially work on the subtitles that find their way into the streamed videos or are unofficially tolerated and/or welcomed to do so.

In June 2007, only a year after the first non-lesbian Web series reached international media attention (both online and offline) (Kuhn 51), the first Spanish Lesbian Web Series, *Chica Busca Chica*,[6] and the first US Lesbian Web Series, *Girltrash!*, both emerged. The medium is vivid and unsteady, meaning the availability of videos may shift daily because of content bans, possible copyright issues, or financial reasons. This ephemeral nature of Web series paradoxically encourages the audience to engage in their stories even more because, unsurprisingly, they want to watch their favourite show before it might be taken off the Internet. Moreover, the audience constantly has to keep track of when and where a new Lesbian Web Series streams its episodes online. Since this development requires a strong viewer commitment to researching the desired websites on time, an online community quickly starts to build, organizing itself through quick media by passing on information regarding the shows through social media such as Facebook or Twitter. This creation of an imagined subcultural community was particularly interesting in the case of scenes taken from the soap opera *Guiding Light*. Two closeted lesbian characters, Olivia and Natalia (known together as "Otalia"), inspired hundreds of fan videos on YouTube that shifted the focus solely to their storyline and, through creative intercuts, made Olivia and Natalia protagonists in videos exclusively dedicated to them. Thanks to these efforts by their fans, Otalia became

---

5 The professionalism of these "translators" has not been deeply investigated so far and would need further research; hence the quotation marks.
6 In English, *Girl Seeks Girl*.

internationally known on YouTube, providing the perfect fan base for their Lesbian Web Series, *Venice the Series*. Bridging the gap between the famous soap opera couple and the new Web series couple was not so hard either since both actresses[7] were the same. Through numerous references and similarities in character creation, music, and clothes, *Venice the Series* (2009–) managed to engage a significant number of Otalia fans from *Guiding Light*.

Participatory Cultures through Quick Media

Web series encourage the emergence of participatory cultures, which, according to Henry Jenkins, is "a culture with relatively low barriers to artistic expression and civic engagement, strong support for creating and sharing creations, and some type of informal mentorship ... In a participatory culture, members also believe their contributions matter and feel some degree of social connection with one another" (*Confronting the Challenges* xi).

The following timeline (which only includes a selection[8] of the very first Lesbian Web Series) demonstrates the connection between the emergence of quick media and the emergence of Lesbian Web Series; it is therefore necessary to always consider the mechanisms of participatory cultures when analysing Lesbian Web Series. The indicated years mark the first appearance of each medium or series: 2004, Facebook; 2005, YouTube; 2006, Twitter; 2007, Tumblr; 2007, first Lesbian Web Series *Chica Busca Chica* (Spain) and *Girltrash!* (US); 2008, *Anyone But Me* (US); 2009, *Venice the Series* (US); and 2010, *Out With Dad* (Canada). I argue that this succession is not coincidental but that the quick media helped pave the way for the primary formation of Lesbian Web Series. It can clearly be seen how quickly social networks and other online media helped create online communities, which then in turn have a considerable impact on Web series. Facebook, as one of the most important social network platforms worldwide, is constantly expanding, having crossed the one billion user mark in 2012 (Blodget) and connecting one-seventh

---

7 Olivia/Gina played by Crystal Chappell, Natalia/Ani played by Jessica Leccia.

8 Admittedly, the selection of Lesbian Web Series is dominated by Caucasian protagonists, neglecting series depicting other ethnic minorities. However, the mentioned series are well-known and received webfest awards. Since, according to intersectionality theories, other series would require us to take at least one additional axis of difference (ethnicity) into consideration, they need further investigation.

of today's world population (cf. US Census Bureau). Taking this fact into consideration, the representation of sexual orientation in quick and other online media – either on Facebook, Twitter, and YouTube or in Web series broadcast on other platforms – is significant for lesbian women, especially in terms of heightening visibility and community building.

Despite the fact that lesbian[9] women have been heavily involved in political activism since the 1970s, they continue to lack visibility in popular culture because of restrictive cultural codes. Admittedly, TV shows integrate lesbian and/or bisexual characters more frequently nowadays (though primarily on private networks), but these characters generally have a secondary function to the narrative and stereotypical depictions often remain. If we want to take into consideration fictional lesbian protagonists in their full complexity within the transmedia context of Lesbian Web Series, we have to dig for a deeper meaning of representation than simple on-screen visibility. Lesbian Web Series focused on fictional lesbian protagonists long before television (with the exception of *The L Word*). For example, *The Fosters* was introduced as a lesbian-centred family show on ABC Family only in 2013, and Ryan Murphy developed his lesbian-centred drama *Open* in 2013–14 before HBO decided not to take the show any further than the pilot.

As much as Julianne Pidduck might want us to believe that "hyper-visibility" regarding lesbian representation already prevails on our screens, the fact remains that "homosexuality" as a disease was only removed from the World Health Organization's International Classification of Diseases in 1990 (cf. Careaga and Curzi 10). Cultural changes in European and North American societies are realistically following, but at a much slower pace than that suggested by any analysis of "hypervisibility" based on, for example, a vivid, flourishing LGBTQ film (festival) scene and movie production. Especially in times of highly discussed anti-gay propaganda laws in Russia and suicide attempts of American homosexual youth in the media (visibility certainly, but of a macabre kind), projects like ItGetsBetter.org – which tries to promote LGBT people's visibility transnationally by collecting videos with encouraging

---

9 To avoid any misunderstandings or omissions, in the words of ILGA (International Lesbian, Gay, Bisexual, Trans and Intersex Association) Women's Project Coordinator Patricia Curzi, the term "lesbian" "refers to any person who identifies herself as a lesbian, bisexual, butch, femme, androgyn, dyke, trans, queer or does not wish to be identified at all with any of the existing terms" (Careaga and Curzi 5).

messages about homosexual identities – prove the need for and importance of a certain reassuring visibility, which does not deserve the prefix "hyper" quite yet. Moreover, in light of research such as "GLAAD Study Reveals Severe Lack of LGBT Characters in Hollywood Films" (McCormick) and "Survey by GLAAD Shows Drop of LGBT Representation on US Television" (Roberts), the discourse on visibility is almost automatically reignited. Furthermore, Lesbian Web Series benefit from using all kinds of different online networks and quick media, connecting women worldwide and making the lesbian communities visible (though not hypervisible yet). As a consequence, a subcultural kinship among (lesbian) women is not only flourishing online but also circulating offline when women start to step out of their "Internet corset" to breathe in next to an immediate subcultural kin, possibly at a fan event.

Lesbian Web Series Reshaping the Perception of Homosexuality, Creating Subcultural Kinship

Apart from the question of visibility, Lesbian Web Series are also significant for the process of identity construction and the emergence of a subcultural kinship. In addition to serial production, fictionality, and narrativity, authenticity is a typical Web-series-related instrument to increase the process of identification between a fictional character and the audience; this leads to an interaction between the fluid groups of makers and consumers (Kuhn 54–3, 58). Regarding the Lesbian Web Series' production, this process of identification plays a key element in order to build up and keep an audience. Keeping in mind what the Bechdel Test has shown us about how predominantly males are represented in the film and TV industries, Lesbian Web Series, in contrast, explicitly focus on (lesbian) women. This constellation automatically reduces female representations in relation to men and at the same time questions the traditional concept of family (father, mother, and their biological children). Furthermore, Lesbian Web Series deconstruct, reconstruct, and replace this one-dimensional concept by approaching the audience with a diverse variety of family constellations, among them a certain demonstration of kinship that is based not necessarily on blood relations but on common cultural assets. In order for this kinship to develop among the predominantly female fans of Lesbian Web Series, "media convergence" can be detected in various ways: according to Jenkins, it comprises "the flow of content across multiple media platforms, the cooperation between multiple media industries, and the migratory behaviour of media audiences who will go almost anywhere in search of the kinds of entertainment experiences they want" (*Convergence Culture* 2–3). All three of these factors play an important role in analysing the material taken from YouTube because these videos reach a more international audience than local TV broadcasts, and moreover strongly contribute to the identification of the audience with the main female characters. According to Mikos (63–4), this identification of a spectator with a certain character depends not only on personal background and beliefs but also on the function of this character embedded in the narrative as a whole. Moreover, the production and perception also depend on the socially circulating moral and ethical codes within society. In order to identify with a lesbian protagonist, viewers have to overcome or at least put aside the hegemonic heteronormative moral codes dominating most societies.

Lesbian Web Series denaturalize heteronormativity (but not automatically heterosexuality) on-screen. As Berlant and Warner argue, "the institutions, structures of understanding, and practical orientations that make heterosexuality seem not only ... organized as a sexuality – but also privileged" constitute heteronormativity as a social norm within society. They further remark that "[i]ts ... privilege can take several (sometimes contradictory) forms: unmarked, as the basic idiom of the personal and the social; or marked as a natural state; or projected as an ideal or moral accomplishment" (548). In consequence, Lesbian Web Series, as a transnationally emerging and constantly developing field across multiple platforms, not only challenge heteronormative standards but also invite a different approach to the concept of family. They depict a new form of online culture by reshaping homosexuality as one of many possible social and cultural standards, not as an anomaly. As Lisa Lowe remarks, culture has often been misinterpreted as a "static repertoire of symbols" (39), while on the contrary it is a complex, oscillating entity that constantly constructs and deconstructs itself at the same time. In the case of Lesbian Web Series, they mainly seem to function within their own participatory culture. The borders between filmmakers and the (in this case, predominantly female) audience get blurred, allowing the individual spectator to influence the active process of creating a Web series and thus also the formerly supposed passive role of watching.

At the same time, a general shift in the entertainment industry regarding female characters and the concept of family can be detected, particularly among Web series. In April 2013, for instance, The Dinah (the Dinah Shore Weekend), a lesbian event in Palm Springs, introduced a film night for the first time, screening the new LGBT Web series[10] *Second Shot* (2013–) and a documentary on the festival itself. In 2014, there was an even larger screening of lesbian-oriented media planned, and the film night scheduled for 2015 made it a tradition. The Dinah is one of the best-known international lesbian festivals, and women from diverse parts of the world participate in celebrating their identity as lesbian women together, watching Lesbian Web Series screenings, listening to concerts with lesbian content, or participating in Q&A panels with famous women within the lesbian subculture (e.g., comedians, actresses). The level of exchange on norm-challenging identity or family concepts within this transnational lesbian subculture during such festivals should not be underestimated. The mutual encouragement of female-centred discourses allows participants to develop strength and self-confidence for the hostile surroundings they sometimes have to face outside of safe places like these events. Provoking a shift in female characters and family concepts in media representations is an "act of resistance" and therefore not always an easy task. As Fiona Joy Green and May Friedman mention in their introduction to *Chasing Rainbows: Exploring Gender Fluid Parenting Practices*, "acts of resistance can take place in small, intimate settings through private conversations in person, over the phone and on line, as well as in larger and more public activities such as parades, performances, camps or media events" (9). These acts of resistance against common social norms – whether they are about gender-fluid parenting or diverse fictional lesbian protagonists in transmedia narratives – are a crucial element in the process of reshaping female representations as well as in reshaping the perception of this very representation. Lesbian Web Series clearly put their emphasis on gay women, who constitute the main characters and around whom their stories unfold. Nevertheless, the last couple of years on US TV have also shown that male writers and producers are slowly starting to challenge gender roles[11] in their shows as well.

---

10 The actress Jill Bennett calls it a "sitcom" (a genre of comedy), although on the promotional shot glasses distributed during the Dinah it is called a "web series," which also seems appropriate since it refers to its production format and online streaming.

11 Numerous shows have depicted female lead characters over the past few years, such as the Fox series *Fringe* (2008–13), which showed the female protagonist as a strong and emotionally complex FBI agent in action, while the two male characters were portrayed as a talkative father and son, cooking together and working out their relationship.

Interestingly, as soon as (straight) women in media narratives are portrayed as strong, confident, and self-sufficient, the audience immediately speculates about the character's (and often times even the actress's) sexual orientation, assuming she "must be" gay. The following YouTube comment illustrates such an interest in "real-life lesbians" by the audience, as ElFrankus asks, "'Stupid' question: Are these girls really lesbians or just (precocious, amateur) actresses?" (OutWithDad). The Bechdel Test has proven to be a useful tool to filter female characters that do not exclusively exist in relation to men. Seemingly, this comes more naturally to lesbian protagonists as they are not expected to fulfil any predefined role in relation to men. Precisely because Lesbian Web Series constantly question discursive constructions of the female self and look for new ways and means to (re)create it, they raise issues about womanhood and nationhood in ways that this collection frames as a "transnational sensibility" (Friedman and Schultermandl 5). Launching such a Web series often generates a critical discourse on identity constructions within its transmedia context. Even though hybridized families (including "rainbow families" of same-sex couples) and their diverse realities are taken into consideration, the debates about supposedly stereotypical lesbian haircuts and clothing or a missing ethnic representation once more point out the difficulties for filmmakers of doing justice to non-normative female representations. However, living a concept of diversity while questioning norms and social conventions is not a privilege for lesbians on and off the screen alone. Queerness can only operate within a context of norms, breaking and deconstructing them until – in its purest (hypothetical) form – it leads to a deconstruction of itself, its own system, or theory.

Fan Engagement and Exchange through Quick Media

Arguably, the starting points for imagined, predominantly female, transnational (online) communities and subcultural kinship are rooted in lively engagement and exchange within the series' context, even in its smallest form. Moreover, the Web allows for a family, no matter in what kind of constellation, to consciously choose what to watch with their children without the usual temporal restrictions of television shows, opening up a new way of accessing representations of non-traditional families. Admittedly, there are recently introduced offerings such as *The Fosters* (2013), which aired at 9:00 p.m. on ABC Family. This series – built around a lesbian couple and their biological, adopted, and foster children – also tries to offer a new approach by representing a diverse, multi-ethnic family (after *The L Word* [2004–9]) but the established (late) broadcasting time remains.

The recent phenomenon of online Web series differs from the familiar format of TV series in three significant ways. First, as mentioned before, the audience has the freedom to choose when they watch episodes online, and filmmakers do not have to follow the usual restrictions linked to broadcasting time on TV. This is especially important when talking about those shows which portray lesbian women as main characters. Representations of lesbian women on daytime television, when they occur at all, often lack realistic depictions of physical intimacy, as lovers Olivia and Natalia on *Guiding Light* demonstrated with constant hand-holding and touching foreheads instead of kissing, cuddling, or any kind of suggestion of sexual activity. Second, filmmakers need to make sure their series actually finds its way to their fans' TV or computer screens. Web series creators cannot rely on users zapping between TV channels and thus accidentally getting to know a show or using a TV guide for program navigation, and so they must use different business models (crowdfunding, merchandise sales, etc.) in order to promote themselves and their art. Last, the Internet provides an immediate and broad transnational access to Lesbian Web Series in numerous countries at the same time.

However, the Web series medium not only plays an essential role in how lesbian identities are narrated, but also effects the development of its broader context, including websites, fan fiction, and poster competitions. According to Henry Jenkins ("Transmedia Storytelling 101") and Nuno Bernardo, these contexts form part of "transmedia storytelling," which crosses the boundaries of different media in which the stories circulate. Additional or alternative storylines are told through quick media, in forums, and on websites. Sometimes, they are taken up or even asked for by the producers or screenwriters and, in turn, incorporated into the storyline. To involve the target audience even more in their series, the creative minds behind these shows frequently encourage their fans to participate in DVD cover or poster competitions. Thus, Lesbian Web Series filmmakers and their audience can no longer be considered as separate, oppositional groups but instead are one community, which creates a new form of cultural assets and a participatory culture that "shifts the focus of literacy from individual expression to community involvement" (Jenkins, *Confronting the Challenges* xiii). Therefore, a heavily female-oriented, transnational community is shaping and reshaping itself constantly, thriving within a hybrid linguistic and cultural space which itself is highly versatile.

On this transnational level, the European group of the International Lesbian, Gay, Bisexual, Trans and Intersex Association (ILGA-Europe) reinforces the "[m]ainstreaming of LGBTI/rainbow families in formal education, and public information materials thus addressing invisibility and promoting the notion that all families have equal value." Furthermore, they promote "[v]isibility of LGBTI/rainbow families (with or without children)" and call for them to "be included in the media and elsewhere in society among other family forms" (ILGA-Europe). Despite the fact that Lesbian Web Series can be seen as in favour of ILGA-Europe's key demands, encouraging a discourse on the visibility of diverse female identities and rainbow families, it cannot be ignored that the practical implementation of the ILGA's idea – e.g., regular funding for queer cultural production – is still missing.

Transnational Bonding, Identity Reinforcement, and Community Building through YouTube Comments among Lesbian Web Series Fans

Participatory cultures throughout the Web are erasing borders between filmmakers and their audience: fans who watch Lesbian Web Series are repeatedly artistically inspired by their favourite shows, supporting them with fan videos and blogs or paintings featuring the main characters of the individual shows. Admittedly, one might argue that it can be considered common fan behaviour to create a fan community. However, in the case of Lesbian Web Series there are two facts worth mentioning: first, most of the fans who actively interact with the Web series makers seem to be women; and second, the bonding occurs not only among fans who share opinions on the show but among those who repeatedly mention their thoughts on either their own identity or that of someone else close to them. The focus hereby often lies on the fan's sexual orientation along with the strongly linked question of belonging to someone or some sort of kin in a broader, non-blood-related, sense.

Comments posted on YouTube[12] – where the Canadian *Out With Dad* (2010–), the American-based *Venice the Series* (2009–), and the Spanish-based *Chica Busca Chica* (2007) air their shows and where their videos

---

12 Comments on other platforms or websites are not taken into consideration in this article in order to guarantee a representative comparison of the three series' related comments.

are reposted by fans – indicate where cultural kinship through Lesbian Web Series is born. As numerous posts demonstrate, among the female audience many viewers identify as lesbian (dalk7293: "I am a lesbian"; OutWithDad) or bisexual themselves (croladyjulja: "i [sic] am bisexual ♥"; OutWithDad) or give lesbianism a positive connotation like, for example, yupi89: "viva el lesbinismo [sic] viva [E]spaña!!"[13] (Natiuxx). To feel a sense of belonging by building an imagined family through quick media communication is even believed to save lives. Lesbian Web Series, as commenter renville68 stresses, can function as a tool to help build and strengthen this community:

> For those that think that this kind of storyline is no longer required, I totally disagree. There are many kids around the world who have no one to talk to about their issues, either because of religion or they live in a small community, and their local tv network won't show this kind of storyline. And the only place they can access info is on the web. In my opinion, these kind of stories can save lives for kids trying to come to terms with their feelings for someone of the same sex. (OutWithDad)

While watching the series and their lesbian protagonists, an identification process, especially among LGBT people, can take place as viewers recognize that they are not alone; as AnastaTube remarks, "Their story is like my story :o I am not the only one yeah!" (OutWithDad). During this process of identification, other audience members even add terms like "proud" or "pride" to their prior statements on identity. Some viewers strongly relate to an episode's content, as MickeyDs14 shows by posting,

> I don't think they're overthinking it. I'm a lesbian and I kind of went through this. I'd never been with a girl before so I thought I just needed to have sex with a guy and when I did it really didn't fulfill anything. I kind of over analyzed everything about myself. Then when I kissed a woman for the first time it was magic. At least they're discussing it, as opposed to being some "bi-curious" girls who are just going through a phase and de-valuing what being gay is. So I kind of understand it. (OutWithDad)

Declarations on sexual orientation and lesbian identity also function as a justification for liking a show, as the user Fabiola Magini demonstrates

---

13   "Long live lesbianism, long live Spain!!" (author's translation).

when commenting on *Chica Busca Chica*: "ME GUSTA MUCHO ... YO SOY LESBIANA ... BESOS PARA TODOS [*sic*] LAS CHICAS!!!!!"[14] (Natiuxx). These open declarations on sexual orientation, identity, and feelings towards the same sex within such a public and internationally accessible medium create a discourse that, without the Internet, quick media, and specifically platforms like YouTube, would not have been able to develop in such an immediate way in the first place. Patricia Curzi, the Women's Project coordinator at the ILGA, points to how important it is for a diverse cultural production to, on the one hand, generate lesbian narratives (other than porn) and on the other to have an audience respond and discuss them: "Sharing experiences and knowledge is a way to develop skills and being aware of those achievements is the first step towards empowerment and pride" (Careaga and Curzi 71). However, what matters is not only the creation of such narratives and visibility but also portraying lesbian couples authentically. *Venice the Series* and its actresses show what was mostly censored on TV in *Guiding Light* – for example, a simple and passionate kiss between two women lying in bed. The Web series demonstrates that no nudity is necessary to get the message across to the audience, which reacted with numerous YouTube comments:

kshatrani7: These kisses are what they never did in Guiding Light!

Chobits827: When these actresses played Olivia and Natalia their characters rarely ever kissed. It almost seemed like they were good friends and not a love interest. I always wondered if it was do [*sic*] to one of them being uncomfortable touching another women in that way. So seeing the opening of this series [Venice] shocked the heck out of me!

kelsi reazin: @Chobits827 ya the only reason otalia didn't kiss is cuz of fn [*sic*] cbs network, cc [Crystal Chappell] n jl [Jessica Leccia] even tried to push for kisses.

RosaLeeAndMe: OOOOO The women kissing is what our show need i [*sic*] think. (Venicetheseries)

---

14   "I really like it [the show] ... I am gay ... kisses to all the girls!!!!!" (author's translation).

As kelsi reazin points out in her entry, the actresses even tried to deliver intimate scenes on the soap opera in a more realistic way, but those scenes never made it through the final cut. That is why *Venice the Series* and quick media used in a transmedia context are a good example for the freedom of expression that the Internet[15] and all its tools have to offer. Authenticity, as one of the essential characteristics of Web series mentioned earlier, is crucial for a lesbian kinship to develop empowerment and pride. Before the first appearance of Lesbian Web Series in 2007, broadcasting *The L Word* on Showtime stirred up a critical academic discourse (e.g., *Reading the L Word: Outing Contemporary Television*, Akass and McCabe) on what authentic lesbian identity representation in media should preferably look like: a diverse group regarding gender, sexual orientation, ethnic minorities, and disabilities. Despite the fact that it is nearly impossible and admittedly unrealistic to represent women in such a colourful way in only one show, it is no surprise that among the huge variety of emerging Lesbian Web Series, most of them (try to) cover these different "required" aspects in representation or exclusively cover an otherwise neglected one by integrating characters of diverse ethnicities or any other diverse background (religious upbringing, etc.). In a comment on *Chica Busca Chica*, viewer meggitan expresses doubts that the series tries to represent such a broad diversity of the "lesbiana española" (the "Spanish lesbian," as she calls her) and additionally remarks that, for her, the protagonists just "happen to be gay" and that the show is more about different personalities than sexual orientation. Nonetheless, it cannot be left unstated that the transmedia context in which these Lesbian Web Series circulate and operate is strongly (although not exclusively) built on sexual orientation and the "lesbian identity factor." Lesbian-themed (online and print) magazines keep the conversation going, by publishing recaps of shows regularly, taking audience polls on the best lesbian actresses, or providing links to the latest video of GLAAD[16] awards. Furthermore, telling lesbian-centred stories on one Lesbian Web Series not only provokes discussions but also seems to encourage filmmakers

---

15   The Internet is not free of censorship and certain restrictions, of course, but in the case of LGBT identities it offers a greater freedom regarding the depiction of diversity and authenticity.

16   The Gay & Lesbian Alliance Against Defamation (GLAAD) monitors and creates LGBT identities in media and reports discrimination against LGBT people. For more details go to www.glaad.org.

within the medium to create new ones; Nadine Bell from Canada, for instance, was inspired to create *LESliVILLE Series* (2013) after watching Crystal Chappell's *Venice the Series*. Bell stresses the importance of a lesbian online community closely connected to lesbian on-screen representations, on the Web as well as on television:

> Though I only ever lurked in the shadowy corners of fandom I was raised from a wee baby dyke to a full grown lesbian woman by that online lesbian community. Fanfic writers turned original fiction authors, those wonderful souls that uploaded clips and videos of gay characters from shows and movies I had NO access to otherwise (the days before YouTube), and the forums and fan sites that supplied all the images and screencaps a gay girl could ask for (the days before Tumblr), all of these anonymous online beings gave me the stories and stereotypes, the romantic cliches, and cultural history that I knew were missing from my offline life. (Bell)

Bell not only actively participated in the lesbian online community as an attentive audience member, but she also worked closely with Jason Leaver on his show *Out With Dad* and eventually launched her own Canadian Web series, *LESlieVILLE Series*, at the beginning of 2013: "Many years later, after a healthy dose of life experience, I wanted to give something back. I wanted to write a story and film it and share it online with the community that gave me so much for so many years" (Bell). She is a perfect example of an audience member feeling supported by this female-centred online subculture formed around Lesbian Web Series while at the same time strongly engaging in it.

To conclude, the present phenomenon of Lesbian Web Series emerging online encourages the viewers' identification processes with the characters but also challenges conventional ideas of the self while shaping a transnational discourse on lesbian representation in media. The widespread "transmedia storytelling," in which the Lesbian Web Series are embedded across multiple platforms and communicated through quick media, poses a particular challenge for scholars. Web series and websites are constantly (de)constructed and therefore represent a continuously transforming and shifting corpus that nurtures the discourse to be investigated. Despite the fact that the Spanish *Chica Busca Chica* was one of the first two Lesbian Web Series to have ever aired online, the anglophone dominance of more recent Web series showing lesbian main characters has to be seen critically through the lens of subcultural lesbian kinship. It remains to be seen what aspects of non-anglophone audiences of Lesbian Web Series are affected by the imposed cultural hierarchies, as well as if and how the Web series production processes themselves are influenced by the interactions with their transnational audiences.

Moreover, the diversity of aspects which circulate within Lesbian Web Series themselves range from new power relations and queered hierarchies to a multiplicity of representations of extended concepts of family. The latter offer an additional "home" to audiences who look for something other than heteronormative narratives within the World Wide Web and who establish cultural kinship by sharing similar values and beliefs through quick media. Not to be misinterpreted, the act of queering "Web" families, of celebrating lesbian sisterhood, and of creating communities outside the heteronormative narratives which usually dictate the mainstream media, shapes new values and cultural assets that do not, by any means, form part of "global feminism" (cf. Kaplan, Alarcón, and Moallem 7) or create a global lesbian subculture. The flourishing transnational kinship among lesbian women remains, however.

Lesbian Web Series and TV Shows with Significant Lesbian Characters

Anyone But Me (2008–12)
Chica Busca Chica (2007)
Girltrash! (2007–9)
Guiding Light (1937/52–2009)
LESliVILLE Series (2013–)
Open (HBO cancelled the show in 2014 before airing it)
Out With Dad (2010–)
Second Shot (2013–)
The Fosters (2013–)
The L Word (2004–9)
Venice the Series (2009–)

WORKS CITED

Akass, Kim, and Janet McCabe. *Reading the L Word: Outing Contemporary Television*. London: I.B. Tauris, 2006. Print.
Alarcón, Norma, Caren Kaplan, and Minoo Moallem. "Introduction: Between Woman and Nation." *Between Woman and Nation: Nationalisms, Transnational*

Feminisms and the State. Ed. Caren Kaplan, Norma Alarcón, and Minoo Moallem. Durham, NC: Duke University Press, 1999. 1–16. Print.

Bell, Nadine. "About LESlieVILLE." http://llvseries.com/about-3/. Web. 14 Oct. 2013, no longer available online.

Berlant, Lauren, and Michael Warner. "Sex in Public." *Critical Inquiry* 24.2 (1998): 547–66. Print.

Bernardo, Nuno. *The Producer's Guide to Transmedia: How to Develop, Fund, Produce and Distribute Compelling Stories across Multiple Platforms.* Lissabon/ Dublin/London: beAktive books, 2011.Print.

Blodget, Henry. "Well, Now We Know What Facebook's Worth – And It's Not $100 Billion." *Business Insider.* http://www.businessinsider.com/how-much-is-facebook-worth-2012-2. Web. 24 Apr. 2012.

Cardini, Daniela. "Looking for a 'Place in the Sun'? The Italian Way to Soap Opera." *Soap Operas and Telenovelas in the Digital Age: Global Industries and New Audiences.* Ed. Diana Rios and Mari Castañeda. New York: Peter Lang, 2011. 111–28. Print.

Careaga, Gloria, and Patricia Curzi. "Lesbian Movements: Ruptures and Alliances." *International Lesbian, Gay, Bisexual, Trans and Intersex Association (ILGA),* Brussels, 71. http://ilga.org/ilga/en/article/lYwN1bs14T. Web. 28 Mar. 2013, no longer available online.

dieStandard.at editors. "Kinos führen 'Bechdel-Test' ein." *dieStandard.* diestandard.at/1381371510436/Schweden-Kinos-fuehren-Bechdel-Test-ein?ref=article. Web. 13 Nov. 2013.

Friedman, May, and Silvia Schultermandl. "Introduction." *Growing Up Transnational: Identity and Kinship in a Global Era.* Ed. May Friedman and Silvia Schultermandl. Toronto: University of Toronto Press, 2011. 3–18. Print.

Green, Fiona Joy, and May Friedman. *Chasing Rainbows: Exploring Gender Fluid Parenting Practices.* Bradford, ON: Demeter Press, 2013. Print.

International Lesbian, Gay, Bisexual, Trans and Intersex Association (ILGA-Europe). "ILGA-Europe's Key Demands for the Recognition of Diverse Families." ILGA. 2012–13. http://ilga-europe.org/resources/policy-papers/family-policy. Web. 11 Oct. 2013.

Jenkins, Henry. *Convergence Culture: Where Old and New Media Collide.* New York: New York University Press, 2006. Print.

– "Transmedia Storytelling 101." *Confessions of an Aca-Fan* 22 Mar. 2007. http://henryjenkins.org/2007/03/transmedia_storytelling_101.html. Web. 11 Aug. 2012.

– *Confronting the Challenges of Participatory Culture: Media Education for the 21st Century.* Cambridge, MA: MIT Press, 2009. Print.

Kuhn, Markus. "Zwischen Kunst, Kommerz und Lokalkolorit: Der Einfluss der Medienumgebung auf die narrative Struktur von Webserien und Ansätze zu einer Klassifizierung." *Narrative Genres im Internet: Theoretische Bezugsrahmen, Mediengattungstypologie und Funktionen.* Ed. Ansgar Nünning et al. Trier, Germany: WVT, 2012. 51–92. Print.

Lowe, Lisa. "Metaphors of Globalization." *Contact Spaces of American Culture: Localizing Global Phenomena.* Ed. Petra Eckhard, Klaus Rieser, and Silvia Schultermandl. Vienna: LIT Verlag, 2012. 23–51. Print.

McCormick, Joseph Patrick. "GLAAD Study Reveals Severe Lack of LGBT Characters in Hollywood Films." *PinkNews.* http://www.pinknews. co.uk/2013/08/21/glaad-study-reveals-severe-lack-of-lgbt-characters-in-hollywood-films/. Web. 21 Aug. 2013.

Mikos, Lothar. *Film-und Fernsehanalyse.* Konstanz: UVK, 2008.

Natiuxx. "WEB SERIE LESBIANA: Chica busca chica. Episodio 1. Español." https://www.youtube.com/watch?v=8P7G3wZPD-8. Web. 14 Oct. 2013.

OutWithDad. "'Rose with Vanessa' – Episode 1x01: Out With Dad." https://www.youtube.com/watch?v=d4OdATe76q4&list=PLUNCZQ-ITZ49ZnVKJi4M2-mXk9K7UVpmu&feature=player_detailpage. Web. 14 Oct. 2013.

Pidduck, Julianne. "After 1980: Margins and Mainstreams." *Now You See It: Studies in Lesbian and Gay Film.* Ed. Richard Dyer and Julianne Pidduck. New York: Routledge, 2003. 265–94. Print.

Roberts, Scott. "Survey by GLAAD Shows Drop of LGBT Representation on US Television." *PinkNews.* http://www.pinknews.co.uk/2013/10/11/survey-by-glaad-shows-drop-in-lgbt-representation-on-us-television/. Web. 11 Oct. 2013.

Sarkeesian, Anita. "The Bechdel Test for Women in Movies." https://www.youtube.com/watch?v=bLF6sAAMb4s. Web. 29 Mar. 2013.

Sarkeesian, Anita. "The Oscars and the Bechdel Test." https://www.youtube.com/watch?v=PH8JuizIXw8. Web. 29 Mar. 2013.

US Census Bureau. "World POPClock Projection." http://www.census.gov/population/popclockworld.html. Web. 24 Apr. 2012.

Venicetheseries. "Venice the Series – Web Series Episode 1 Season 1." https://www.youtube.com/watch?v=f43ZzXt4EKA&index=1&list=PL73BE9C5145 40C2B8. Web. 14 Oct. 2013.

5 Literary Letters and IMs: American Epistolary Novels as Regulatory Fictions

SILVIA SCHULTERMANDL

In *Simians, Cyborgs, and Women: The Reinvention of Nature* (1991), Donna Haraway argues that in every historical moment of Western modernity, legal, scientific, literary, and religious discourses shape concepts of gender. These discourses, which both describe and prescribe gender norms, circulate in the form of narratives about the gendered body. Because of their ideological power of shaping social structures, Haraway calls these narratives "regulatory fictions" (135). In Haraway's own project, regulatory fictions refer to print texts whose dissemination was facilitated through two phenomena beginning in eighteenth-century Europe: increased literacy and the rise of the printing presses. In the context of quick media, these regulatory fictions come into effect through participatory practices of users in online communities. Kerry Mallan, for instance, argues that "online communities produce a range of discursive practices and expectations, which attempt to constitute young people in particular ways" (66). Mallan's extension of Haraway's concept of regulatory fictions relates to empirical work on adolescent users of online communication devices and their representation in young adult literature. The network behaviour Mallan analyses in both real and represented contexts of quick media usage by adolescents sheds light on the particular imagination of space, power, and knowledge in online communities. In contrast, my analysis of examples of American epistolary fiction traces the epistemological function of regulatory fictions to late-eighteenth-century American nationalism, when a white, Protestant, heteronormative society became dominant in part because of the proliferation of literary texts which propagated nationalist values. To this end, I examine the negotiation of national identity in American epistolary fiction by looking at the discursive shifts from letter writing in eighteenth-century novels to instant messaging in contemporary adolescent novels. The examples I discuss here are William Hill Brown's *The Power of Sympathy* (1789) and Lauren Myracle's adolescent novel *ttyl* (2004). Both novels address young reading audiences and treat issues of femininity and female morality as conscripted by a US nationalist doctrine. While the novels mimic different media, the messages they profess are astonishingly similar.

This emphasis on national identity was a strategic move for the young American Republic in order to distance itself from its English colonial heritage. In *Transnational Nation: United States History in Global Perspective since 1789* (2007), Ian Tyrrell argues that even during the period of intensified nation building in the nineteenth century, the economic, social, and cultural borders of the United States were permeable. It is perhaps no coincidence that Tyrrell takes as a starting point for his study the year 1789, the year the American Constitution was ratified and the year America's first novel was published. But 1789 marks also the publication of Olaudah Equiano's *Interesting Narrative*, an autobiographical account which records incidents of migration, globalization, and transnationalism. Equiano, possibly a native of South Carolina (Carretta), was born into slavery and thus deprived of the citizenship rights that entitled white Americans to a national identity. In Equiano's text, the nation plays less of a role than does movement between nations, transatlantic social movements, and interconnected histories. The publication of these two texts alone suggests that national and transnational trends were present in late-eighteenth-century America. Nevertheless, the strategic emphasis on national allegory in eighteenth-century literature ultimately led to a fortification of a sense of Americanness which adopts the nation-state as primary representational logic. Transnationalism, which is a social reality relevant in both Brown's and Myracle's respective eras, gets largely sidelined.

Reading (for) Kinship

Reading epistolary fiction as regulatory fictions connects to the ever-recurring question of how literature can function as an instrument of community building. Within reader-response theory, this question is largely framed through the idea of recognition. In *Uses of Literature* (2008), Rita Felski, for instance, discusses recognition as one of four potential aesthetic effects of reading literature, the other three being enchantment, knowledge, and shock. Distinguishing recognition from the more unilateral practice of identification, Felski suggests that "[w]hen we recognize something, we literally 'know it again'; we make sense of what is unfamiliar by fitting it into an existing scheme, linking it to what we already know" (25). This etymological precision of recognition as re-knowing something rather than identifying with something unknown highlights the "metaphorical and self-reflexive dimensions of literary representation" (44) and leads Felski to conclude that "[w]e do not glimpse aspects of ourselves in literary works because these works are repositories for unchanging truths about the human condition ... Rather, any flash of recognition arises from an interplay between texts and the fluctuating beliefs, hopes, and fears of readers, such that the insights gleaned from literary works will vary dramatically across space and time" (46).

Extending Felski's observation on the potential of recognition and applying it to the context of multiculturalist and transnational literatures, Winfried Fluck proposes a notion of reading as a form of transfer between "the narrative of the text and the narrative of the reader" (60). Reading literature, in Fluck's sense, means becoming aware of one's own identity through the encounter of the protagonist's identity so fundamentally different from one's own. Narratives of self-fashioning, where a protagonist is on a metaphoric quest to come to terms with his or her identity, exemplify this process of transfer by incorporating it into the level of narration. By reading about a protagonist's search for identity, readers may relate to the practice of coming to terms with their own identities but not necessarily with the protagonist's identity as such.

In the context of epistolary fiction, this process of recognition is in part obscured through the reader's special position towards the narrative, namely the fact that because the letters always address a fictional reader (i.e., another character) directly, they seem to address the real reader as well. This impression of being addressed in person facilitates the reader's identification with the recipient of the letters and a glimpse of recognition of the epistolary exchange. In William Hill Brown's *The Power of Sympathy* and Lauren Myracle's *ttyl*, this interplay between identification and recognition subsumes the reader into the respective communities of protagonists and their shared cultural, moral, and national values.

As a genre, epistolary fiction is commonly associated with the eighteenth century. The dramatic storylines of novels such as Johann

Wolfgang von Goethe's *The Sorrows of Young Werther* (1774) and Samuel Richardson's *Pamela* (1740) and *Clarissa* (1749), as well as Pierre Choderlos de Laclos's *The Dangerous Liaisons* (1782) and Jean-Jacques Rousseau's *Julie, or the New Heloise* (1761), mesmerized audiences through the simulated intimacy and immediacy of the act of reading letters. Although the epistolary novel's popularity in the late eighteenth century was followed by a swift decline (Altman 3), it celebrated its comeback in the twentieth century through modernist and postmodernist adaptations of the classical epistolary form in novels such as Stefan Zweig's *Letter from an Unknown Woman* (1922), Saul Bellow's *Herzog* (1964), Stephen King's *Carrie* (1974), John Barth's *Letters* (1979), Alice Walker's *The Color Purple* (1982), John Updike's *S* (1988), and Lionel Shriver's *We Need to Talk About Kevin* (2003). With the emergence of quick media as a new form of communication, authors have also begun to appropriate quick media formats for epistolary fiction. Daniel Glattauer's *Love Virtually* (2006), Gary Shteyngart's *Super Sad True Love Story* (2010), and Lynn Coady's *The Antagonist* (2011) embrace the format of email conversations for their novels' discursive representations. Lauren Myracle's Internet Girls series *ttyl* (2004), *ttfn* (2007), *l8r, g8r* (2008), and *yolo* (2014) even goes so far as to adopt the quick media format, layout, and stylistic properties to convey the exchanges among three teenage girls.

Despite the stark thematic differences between these novels, they all employ the epistolary form to mimic actual correspondences (Altman 6) – John Barth even in the form of an absurdist letter conversation to offer a critique of American academia and the discourses of university administration. The epistolary novel's generic particularities mimic the act of reading letters in the sense that what the recipient of a letter in the fictional world reads is also what the novel's audience reads. On the one hand, this facilitates identification of the external reader with the internal reader (Altman 88). On the other hand, the external reader, reading in fact all and not only those letters addressed to him or her, has a fuller understanding of the story which unfolds through the exchange of letters and creates a discrepancy of awareness between reader and protagonist. This deliberate evocation of the act of reading letters serves as the vehicle of aesthetic experience and resonates with Marshall McLuhan's credo that "the medium is the message" (qtd. in Altman 8).

This particular reading experience equips epistolary novels especially well with the potential to become regulatory fictions that teach young women how to be good citizens. Specifically, the didactic lessons directed at the novels' heroines are designed to control the reader's sense of female agency within patriarchal society. In reading regulatory fictions of good womanhood in Brown's epistolary novel *The Power of Sympathy* and Myracle's new-media epistolary novel *ttyl*, we can apply these lessons to our own lives. This sense of recognition with the fictional world of the novels facilitates kinship formation between the readers and the protagonists in the sense that they both actively acknowledge each other as part of the same imagined family. Therefore, the kinship ties among the protagonists – they are friends, confidantes, relatives, bound together by a shared sense of values – extend to the level of the reader, who, in turn, can experience a sense of closeness to the protagonists. It is through this sense of closeness that the didactic message of these regulatory fictions can reach the reader.

Sympathy and Didacticism in the Eighteenth-Century Epistolary Novel

In William Hill Brown's *The Power of Sympathy* (1789), the interplay between identification and recognition is a main agent of the novel's didactic function. Brown's novel is generally heralded to be America's first novel, meaning the first novel published in the postcolonial United States, and has been identified as a key text in shaping national consciousness in the late eighteenth century. The novel itself follows the established pattern of the late-eighteenth-century sentimental novel: the young female heroine, lacking sufficient guardianship in her social encounters with young men, falls in love and displays her availability for courtship by exchanging letters with her suitor. This exchange of love letters signals to readers familiar with sentimental plots that she is morally doomed, which often results in the heroine's death at the end of the novel. In *The Power of Sympathy*, this moment occurs when Harriot writes to her friend Myra about her feelings for Harrington. Brown's novel also incorporates another standard feature of sentimental fiction, namely a symmetrical character constellation: in contrast to the sentimental heroine who gives in to her seducer's advances, her female confidante has a more sober attitude towards romantic love. This is also true for the male protagonists: while the romantic hero gives in to his feelings or intentionally sets out to seduce the heroine and expose her lack of proper values, his male confidante tries to talk sense into him. Already the first letter Harrington writes to his friend Worthy (pun intended and even acknowledged in one of Harrington's letters) displays an exuberance of romantic feelings and is countered by Worthy's sober reply that love at first sight is problematic. This reliance on reason characterizes Worthy as an embodiment of the "image of a middle-class man in his republican virtues and values" (Evans 51).

The binary opposition of reason and sensibility embodied by the cast of young protagonists is framed by older protagonists who serve as their potential guardians. For instance, to the same extent that Worthy embodies reason, Mrs Holmes embodies morality. Her letters relate her experience as a good and proper woman, whose father-in-law, the Reverend Holmes, is often evoked as a social authority. Mrs Holmes's authority, in turn, is validated by the fact that she married the Reverend's son, and lives, after her husband's death, under the guardianship of her in-laws. The convention of a negative example preceding the relationship between the two lovers is usually a case of adultery out of which one of them is born. In *The Power of Sympathy*, this negative example is the adulterous relationship of Harriot's mother Maria with Harrington Sr., a relationship during which Harriot was conceived and which makes her, in turn, Harrington's half-sister. This is one of five internal stories which the protagonists tell each other to exemplify the dangers of sentimentalism. All five in Brown's novel have female protagonists whose tragic endings contain an obvious didactic message.

Didacticism is, in fact, the major function not only of Brown's novel but of the sentimental novel at large. The female heroine and the female reader are presented with explicit warnings about social misconduct. Literature in the early American Republic served as a "conveyor of meaning" (Davidson 73), which, as was the general understanding of the role of the arts in general, was designed to elevate the citizens of the United States and instil in them a sense of community:

> Given both the literary insularity of many novel readers and the increasing popularity of the novel, the new genre necessarily became a form of education, especially for women. Novels allowed for a means of entry into a larger literary and intellectual world and a means of access to social and political events from which many readers (particularly women) would have otherwise been excluded. (Davidson 67)

Similar to the effect of Schiller's *On the Aesthetic Education of Man* (1794) on Europe, Hugh Blair's *Letters of Rhetoric and Belles Lettres* (1783) suggested to American reading audiences that humankind depended on the educative potential of the arts in order to instil in individuals a sense of morality and, by extension, national values. As a new nation, the early American Republic found such new values in the Scottish Commonsense School; American universities like Yale and Harvard devised curricula which "stressed the classics, poetry, eloquence, and the arts as disciplines that would improve moral sense" (Elliott 158–9). For women, who were excluded from formal education, popular literature substituted for the theoretical fields of the American curriculum to address more practical matters, most of all the virtues of republicanist womanhood. This was based on the general assumption that what the new nation needed, in fact depended on, were women who embraced American virtues and would subsequently teach them to their children. Brown's novel, in particular, concentrates on "the necessity of improved education ... with special attention to the need for better female education" (Davidson 134).

The need for literature's didactic influence has to do with the fact that American readers during the early national period had doubts about "the fate of the nation" (White 74). The general feeling of insecurity about the future of America after its emancipation from England demanded constant reification of the nation-building project. As much as the Declaration of Independence (1776) proudly celebrates America's freedom, a strong national identity was still missing. Only with the drafting of the American Constitution (1787) was a sense of national identity attainable. Brown's novel, published the same year that the American Constitution was ratified, addresses these uncertainties about American identity. Naming *The Power of Sympathy* America's first novel (i.e., the first novel published in the United States which was not a pirated copy of an English novel), Brown made a national event out of the publication of his book, linking it to the political sentiment of his era by mimicking the tone and gesture of the preamble to the American Constitution in the novel's paratext. In a similar fashion as how the preamble to the Constitution consolidates a sense of community, in no small part by the opening words "we the people," Brown includes a direct address to the readers in the form of a preamble which records the republicanist virtues of moral education:

> To the young ladies of United Columbia, these volumes, intended to represent the specious causes and to expose the fatal consequences of seduction, to inspire the female mind with a principle of self complacency and to promote the economy of human life, are inscribed, with esteem and sincerity, by their friend and humble servant, The author. (5)

Like other authors of the early American national period, including Benjamin Franklin, Hector St. John de Crèvecoeur, Susanna Rowson, Charles Brockden Brown, and Hannah Webster Foster, Brown emphasizes individualism, self-improvement, and rational thinking (Davidson 134). And like his fellow authors of the sentimental novel, his primary targets were young female readers needing to understand that "seduction is a social disease" (Davidson 182).

The sentimental novel's plot line foresees that the "passive, suffering heroine" (Brey 81) dies at the end of the novel precisely because she lacks guardianship. Her death, expounded in typical sensationalist fashion, is to be read allegorically as a potential fate of the young American nation: if women die prematurely because they lack guidance, they are not able to fulfil their role as mothers of the future Americans. In turn, those women who do live to the age of childbirth literally survived the selective process which ensures that only those (white) women who embrace and embody the virtues of republicanist womanhood are fit to live in the new nation. This conflation of female virtue and American nationalism casts the tragic heroine as anti-model for American reading audiences. Confined through heteronormativity and regulative gender norms, the heroine of the sentimental novel embodies a narrowly defined myth of female identity which mainly serves as a manifestation of white, patriarchal hegemony. This intersection of female and national identity for the sake of solidifying nationalism as well as patriarchy becomes a ready-made stand-in for American identity at large. In the introduction to their critical study on nationalism and concepts of femininity, Norma Alarcón, Caren Kaplan, and Minoo Moallem deconstruct this institutionalization of femininity as a means of nationalism: "Women are both of and not of the nation. Between woman and nation is, perhaps, the space or zone where we can deconstruct these monoliths and render them more historically nuanced and accountable to politics. The figure of 'woman' participates in the imaginary of the nation-state beyond the purview of patriarchies" (12). In this vein, the eighteenth-century sentimental novel exemplifies the dynamics at play in literary manifestations of normative ideals of femininity for the sake of nationalism and pinpoints literature's specific role in the creation, dissemination, and implementation of such ideals.

In the same manner that literature can educate young women, it can also disseminate the "wrong ideas" about womanhood. With the absence of an international copyright law, America was flooded with cheap reprints of English literature, literature which heralded sentimentalism on the one hand and British social norms on the other. Readers during the early American Republic were not only young in the sense that their self-identification as American was still under way, they were also young in the sense that more than half of its citizens were 16 years of age or younger (Mulford xiv). That they read a lot (Winans) and that they read a lot of British literature was a potential threat to the establishing sense of American national identity. In other words, eighteenth-century epistolary novels could only function as regulatory fictions if American readers identified with the American context of the novels.

This would explain why Brown's claim that the story of Harriot and young Harrington is "FOUNDED IN TRUTH" (6, emphasis in original) induces an "illusion of reality and authenticity" (6) as a means of evoking American social realities. In the paratext, *The Power of Sympathy* identifies itself as a cautionary tale through which "the dangerous Consequences of SEDUCTION are exposed, and the Advantages of FEMALE EDUCATION set forth and recommended" (5, original emphases). According to Davidson, "What we have here is another graphic illustration (literally and figuratively) of the role of the printer in the creation of the American novel and in the 'seduction' of the American reading public" (162). At the same time, *The Power of Sympathy* addresses this potentially perilous influence of literature in a letter from Mrs Holmes to Myra where she relates the Reverend Holmes's advice on what young women should read. When asked "[w]hat books would you recommend to put into the hands of my daughter?" (20), the Reverend Holmes stresses the importance of selecting the right kind of reading materials: "Novels, not regulated on the chaste principles of true friendship, rational love, and connubial duty, appear to me totally unfit to form the minds of women, of friends, or of wives" (21).

The epistolary form is particularly conducive to fulfil this didactic function because the reader of the novel may easily identify with the addressee of the individual letters. As Janet Altman explains, the epistolary novel has two readers of the letters: the fictional character and the real-life reader. Reading the close and confidential content of the letters simulates a degree of intimacy between the text and the reader which allows for the reader's identification with the text. Eighteenth-century readers may also have recognized formulas and conventions of letter writing to be the same ones they themselves would use under similar circumstances. Being of the same age and possibly equally intrigued by romantic love, the readers of epistolary novels find moral guidance in the advice the romantic heroine or her confidante receives.

In *The Power of Sympathy*, the guardianship older and wiser protagonists give to younger ones is full of social commentary easily applicable
to the contemporaneous reader. Mrs Holmes, for instance, shares with
Myra her dislike of male sentimentalism, describing how she "cannot
help but smile sometimes, to observe the ridiculous figure of some of
our young gentlemen ... the ridicule of which arises, not so much from
their putting on [a] foreign dress, as from their ignorance or vanity in
pretending to imitate those rules which were designed for an English
nobleman" (54). Such men imitate the style and mannerism of Lord
Chesterfield, an English master of seduction. That the novel's anti-hero
is European, or aspires to pass for a European gentleman, prompts
social commentary among the protagonists about America's colonial
past. In fact, it is not uncommon that the rake is English or French,
often additionally a member of European aristocracy, and thus embodies entirely different values from those held high among the American
middle-class protagonists (Evans). Mrs Holmes states "that most of the
letters I have written to you of late, on female education, are confined
to the subject of study" (93). Her warning to Myra is therefore also
to be understood as a warning issued to the young female readers at
large: "but remember that the knowledge which I wish you to acquire,
is necessary to adorn your many virtues and amiable qualifications"
(94). This also goes for her hope that Myra will be able to "always dis-
tinguish a man of sense from the coxcomb" (90). By this implication, the
reader's social standing as either a model of virtue herself or a potentially fatal victim of romanticism can be seen as foreshadowed by the
novel's plot development.

Epistolary novels follow a rather simple pattern, using predominantly the same plot development, character constellation, and setting,
a fact which makes the reader experience an additional form of recognition. The emotive lives of the protagonists are at times so complex
that they cannot adequately communicate their feelings. When Harriot
writes to Myra and relates the story of Ophelia, a young woman who
committed suicide after an ungraceful seduction, "this unhappy affair
has worked me into a fit of melancholy. I can write no more, I will give
you a few particulars in my next [letter]. It is impossible to behold the
effects of this horrid catastrophe and not be impressed with feelings of
sympathetic sorrow" (40). This exaggerated display of feelings not only
highlights Harriot's sentimental nature but affects the reader's experience of the text as well: with Harriot unable to continue writing, the
narrative stops and leaves the reader with a feeling of incompletion.

Eighteenth-century epistolary fiction establishes kinship ties both on the level of the protagonists and between the protagonists and the readers. National identity being a driving force in the novel's social function, readers and protagonists alike negotiate the fabric of community represented within the fictional world of the novel, as well as the sense of national identity in the early American Republic as a whole. At a time when American national identity was still under construction, and when many competing influences, both national and transnational, informed how Americans felt about their lives in the new republic, novels like Brown's establish a feeling of kinship articulated in the form of a moral imperative for young readers to act responsibly so as to ensure the future prosperity of the nation.

Quick Media, Internet Communities, and Femininity

While contemporary epistolary novels generally surpass the predictable formula of the eighteenth-century sentimental novel, young adult fiction extends the didactic function of sentimental fiction by means of explicitly addressing the dynamics of Internet communities. Interestingly, young adult fiction also represents mechanisms of community building through communication and, typical of the epistolary genre, extends didacticism to the relationship between reader and text. Lauren Myracle's *ttyl*, the first book of her Internet Girls series, shows an astonishing adherence to regulative femininity and treats female coming of age didactically by allowing the reader self-recognition in the text in the same ways *The Power of Sympathy* did for late-eighteenth-century readers. Not only do both novels appropriate forms of epistolary communication as a mode of encoding narratives of female virtue in the face of seduction, they are also novels of self-fashioning and community building which operate with stable and rigid concepts of identity much unlike most poststructuralist and postmodern identity theory. Reducing the complexities of teenage angst to simple binary oppositions, *ttyl* suggests that community building in the online world and in the real world follows the same codes of morality prevalent in late-eighteenth-century sentimental fiction.

With its flashy fuchsia colour and emoticons on the cover, Myracle's *ttyl* evokes the stylistic particularities of contemporary chick lit. Myracle's choice of genre may be considered far from coincidental: Joanna Webb Johnson contests that much of young adult literature specifically positions itself in the vastly successful book market of chick literature, consciously acknowledging that neither "chick lit [nor] its authors ... stake a claim on Great American Novel territory" (142). Myracle's novels in the Internet Girls series hold a particular ground in the tradition of young adult literature, namely the subgenre of "chick lit jr.," a genre which "reflect[s] current language (including slang) and methods of communication that were not present or popular as recently as five years ago" (Johnson 143). Written entirely in the form of short messages exchanged among the three protagonists, *ttyl* mimics the netspeak of instant messaging linguistically and typographically. But *ttyl* does enter Great American Novel territory by consciously embracing the same function as *The Power of Sympathy* had for late-eighteenth-century readers.

Counter to the popular assumption that chick lit is a literary phenomenon related to late-millennial consumer culture, this relationship to the sentimental novel of the eighteenth century connects chick lit to historical forms of women's literature. The consensus among scholars of chick lit, however, is that its historical origin is to be found in the nineteenth-century novel of manners published by such authors as Jane Austen and Edith Wharton. Juliette Wells, for instance, argues that the nineteenth-century novel of manners is the thematic and epistemological basis for much contemporary chick lit: thematically, both incorporate love plots, female maturation, and social mores into their plot development; epistemologically, both are often pitted against what was considered "serious" or "real" literature, the implication being in the nineteenth century that women writers "only" produce texts designed for female audiences and that chick lit is a lesser fictional genre. Like Wells, Stephanie Harzewski draws a connection between chick lit and the nineteenth-century novel of manners; her attention goes not to the question of female literature as a potentially lesser art form but addresses issues raised against the novel as a genre in general. Harzewski thereby draws on examples of eighteenth-century literature, recalling debates as to whether novel reading was a proper pastime activity for women at all. As was the case with *The Power of Sympathy*, the main prejudice against the novel stems from a fear of its influence on young readers who, immersed in the aesthetic illusion of the narrative, may be motivated to imitate the behaviour, language, and customs presented in the fictional world, thus becoming corrupted through the reading experience (Warner). While the epistemological debate over the question of appropriate reading material for women does resonate with contemporary chick lit, Harzewski maintains that the subject matter of the nineteenth-century novel of manners provides the actual historical lineage.

Myracle's *ttyl*, however, is an exception to this tradition precisely because it mimics the intimacy of the eighteenth-century sentimental novel in epistolary fashion and transposes it onto the contemporary world of instant messaging. It also modulates the didactic function of sentimental novels like *The Power of Sympathy* but, because of its gimmicky format, emoticons, and teenage slang, does so without the heavy-handedness of its historical predecessors. What Johnson observes about chick lit jr., namely that "[a] novel can teach without being authoritarian, and its instructional aspect is an important function of the chick-lit.-jr. genre" (146), is particularly true for *ttyl*. Its didactic potential emerges through the reader's identification with the protagonists and the recognition of the processes of community building represented in the novel. Moreover, the feeling of reading somebody's IM (instant message) conversations, produced by the novel's adoption of an IM format, is partly responsible for the novel's attractiveness and success. Myracle's novel taps into the – often bemoaned – phenomenon that millennials generally spend less time reading fiction and more time reading each other's messages. While the novel was first published only in print, the tenth-anniversary edition released in 2014 specifically markets ebook readers or, as the description on Myracle's website specifies, the "iPhone generation" ("ttyl" n.pag.). Reading an ebook novel consisting of texts on one's smartphone connects the act of fictional reading to the ritual of constantly checking one's messages. Reading regulatory fictions packaged in quick media format and delivered through quick media devices perfectly exploits the millennials' communicative behaviour for the novel's didactic function. But the didactic function of Myracle's novel works just as well in the print version, precisely because of the special potential for readers to recognize themselves within the communicative situation of the epistolary novel.

Myracle's *ttyl* follows the lives of three teenage girls – Angela, Maddie, and Zoe – by depicting their intimate and casual IM conversations. Among their regular topics of conversation are bits of gossip about other classmates, frustration about their parents, or objects of consumer culture. Their conversations also display a strong interest in female sexuality, ranging from the more innocent crush on a boy in school to lewd behaviour at a frat party and sexual harassment by a teacher. The novel consists entirely of instant messages and even uses a layout similar to that of IM chat rooms. As already indicated by its title, the novel mimics Internet slang and other idiosyncrasies of the blogosphere, including emoticons and secondary text. The language is colloquial, even profane at times, and, depending on the protagonists' character traits, overtly dramatic. Tension and exhilaration, for instance, are commonly indicated by secondary text, such as in Angela's euphoria that the boy she has a crush on is in her French class: "cuz – drumroll, please – ROB TYLER is in my french class!!! *breathes deeply, with hand to throbbing bosom" (6). The casual tone and dynamic exchange of messages present the reader with a lively community of bloggers.

While the three protagonists encounter similar problems, they themselves are quite different: Angela is romantic, concerned about her looks and appearance, and seems in need of constant external validation for her thoughts and actions; Maddie is rough, vulgar, and streetsmart (to the extent that this is possible for a suburban middle-class teenager); and Zoe is pious, insecure, and proper. These character traits are also amplified by the specific fonts and nicknames in which their IMs appear in the novel: Angela (SnowAngel) uses a curly, blue font which caters to the cliché of girliness in which she is cast; Maddie (mad maddie) writes in a bold, black font implying more assertiveness and a degree of aggression, which corresponds to her behaviour among her friends; and Zoe (zoegirl) writes in a very ordinary, unspectacular font, almost as if she did not customize her writing in order to call less attention to herself.

The novel operates on the assumption that teenage readers will identify with at least one protagonist, no matter their ethnic identity. Perhaps intentionally, *ttyl* never addresses the protagonists' ethnic identities, implying either a post-ethnic attitude towards ethnic diversity or majority discourses. Since all three protagonists are central to the plot development and each one claims as much attention for her individual problem, each of the three serves as the heroine and the confidante at the same time. Culturally, they are largely interchangeable: in numerous instances they each refer to their suburban lifestyles and similar cultural identities. Despite their constant online presence, they are never depicted discussing or engaging in any contact with the online world outside of their close-knit community, much less in the context of transnationalism. Unlike in *The Power of Sympathy*, no national other serves as a foil for cultural distinction. The sheer absence of cultural alterity suggests that the three girls, and perhaps also their primary intended audience, are of the same cultural identity, especially if this identity is solely defined by consumer culture and the rhetoric of choice.

Interestingly, Myracle's novel not only shares the formal characteristics of the eighteenth-century epistolary novel, it also adheres to the same didactic purpose of educating young women about becoming responsible citizens. The main difference lies in the novel's choice to depict peer conversations only: no adults or older correspondents serve as guardians; it is the shared values of the three friends which equip them with the ability to give advice, often unsolicited, about issues of teenage malaise.

The novel implies that suburban American teenage girls are constantly under supervision, either by their parents, their peers in school (often bullies), or their IM friends. This group identity manifests itself in part through a shared sense of oppression on the one hand and centripetal group dynamics on the other. Most of the conversations in the novel are dialogic; group conversations only occur when the three friends discuss plans for a road trip. The dialogues highlight the interpersonal dynamics between two friends, often at the expense of the one who is not privy to the conversation. A large portion of their speech acts is devoted to talking about the third friend, often in a highly critical and not very supportive manner; an equally large portion is communicating their thoughts and feelings the moment they arise. These two topics can change in close succession throughout a longer chat log:

> mad maddie: [Angela] is her own worst enemy, zoe. [...]
> zoegirl: yeah, but it's also just ANNOYING. [...]
> mad maddie: hey, i didn't wanna mention it in front of angela, but can I
>   just tell u what a great time i had at work saturday nite?
> zoegirl: with ian? (84–5)

Alternating between the three characters also makes up the plot structure of several interconnected strands. As in *The Power of Sympathy*, events that are not part of the immediate present of the conversation are only narrated within the IM exchanges in the form of internal stories.

The rapid succession of IM exchange also entails that each protagonist is heroine and confidante at the same time, sometimes even within one chat log. Similarly, the switch from active to passive confidante (Altman 54) signals an important development in the heroine's life. When Maddie feels betrayed because Angela broke a secret she promised to keep, Maddie retreats from the group by turning on an offensive auto-reply message:

> zoegirl: maddie, r u there?
> Auto response from mad maddie: shove it up your ass
> zoegirl: maddie, i need to talk to you, please?
> zoegirl: it's about mr. h.
> zoegirl: he wants to go hot-tubbing with me, and i don't know what to do.
> zoegirl: maddie?
> zoegirl: ok, well, i really could have used your advice, but i guess u don't care! (196)

As a passive confidante who receives the message but refuses to reply, Maddie temporarily interrupts the community building, in turn making the foundations of that community (in her case mutual trust) apparent.

The novel's didactic function is not so much to respect the rules of decency as it is to get teenagers to stay out of "trouble." It cannot be a coincidence that all three, despite their different character traits, end up in compromising situations: Angela finds her boyfriend out on a date with another girl; Maddie gets drunk at a frat party and dances topless, resulting in photos of the incident circulating on the Internet; Zoe mistakes her English teacher's attention for mentoring when in fact he is interested in her sexually, such as when they go hot-tubbing together. While all three incidents end well, the reader is invited to pass judgment on the girls in a similar fashion as they judge each other. Angela's dependency on boys' admiration, Maddie's decision to go to a party with a classmate who is known to cause trouble, and Zoe's inability to detect her teacher's sexual interest are depicted as the result of the protagonists' character flaws and thus reduced from a social to an individual level. The novel does not question the frat party as an institution, teenage co-dependency, or a lack of awareness of sexual harassment; instead, the girls are held accountable for personal choices to put themselves into situations where they may experience embarrassment, shame, or physical and emotional harm. In this sense, *ttyl* picks up the same emphasis on personal responsibility which sets the tone for *The Power of Sympathy,* namely not the general idea that patriarchy and heteronormativity impose limits on young women but rather that women should fear the dangers of sentimentalism and their inability to handle them without moral guidance. SnowAngel, mad maddie, and zoegirl are protagonists located at the intersection of gender and nationhood, as are the readers of *ttyl*, especially in the post-9/11 context of heightened nationalism in the United States. After all, every discussion about immigration, foreign policy, gay marriage, and so forth is always also a rehearsal of nationalist ideals of "an" "American" "identity." What it means to be a good and responsible citizen is as much an issue as it was in the early republic, perhaps with the difference that through quick media outlets, these narratives of the nation-state can reach both national and global audiences and prompt instant reactions.

As is true for the protagonists, the readers of chick lit jr. are presumably also in danger of moral corruption. Framed as a tale to exemplify the dangers of teenage sexual awakening, *ttyl* seems to suggest that the protagonists' friendship and their constant presence in each other's lives via quick media technology provide a metaphoric safety net. Through the reader's potential identification with the protagonists and the flash of recognition the text triggers by mimicking the codes of contemporary American youth culture, *ttyl* extends the realms of kinship established among the protagonists to the readers. The community feeling among readers of Myracle's Internet Girls series even becomes visible through the commercial success of the individual novels and the variety of merchandise they inspired.

Conclusion

Whether written in the form of literary letters or IMs, American epistolary novels exemplify processes of kinship building both among the protagonists and between the protagonists and the readers. In essence, the communication mode does not determine the potential message and function of these novels. Despite the fact that the publication of *The Power of Sympathy* predates that of *ttyl* by more than two centuries, the didacticism the novels depict serves the function of regulating female identity. Both novels contribute to the regulative formation of an imagined community (Anderson) of their respective eras by engaging the reader in a participatory reading experience through selective identification and recognition. Like the participatory mechanisms prevalent in quick media, these practices of identification and recognition are built into the genre itself: through the particular closeness between readers of epistolary fiction and the individual texts, a strong sense of interactivity is able to emerge. This allows, even coerces, the reader to develop kinship ties with the protagonists in a fictional world, a world which they recognize in particular because of the intimate act of reading letters and following the exchange of instant messages.

What is astonishing with these two examples is that national identity is foregrounded to such a large extent. In *The Power of Sympathy*, a strong sense of American national identity is communicated as a means of distancing America further from its colonial past; this emphasis on the national also overwrites the transnational connections that were prevalent in American culture at the time in the form of a rich transatlantic cultural dialogue. By contrast, there is no indication of a transnational connection between the protagonists of *ttyl* and the world around them. In fact, their kinship ties are so exclusive and create the impression of a very insular, even secluded, cyberspace, which is not at all characterized by the global reach of quick media. Nevertheless, both novels emphasize the possibility of a tight-knit sense of kinship and mutual responsibility among the protagonists. By adopting narrative discourses and layouts of epistolarity, both in letter and in IM format, Brown's and Myracle's novels position personal identities as relational to communities which can be accessed through personal correspondence, whether through the practice of letter writing or the use of quick media, respectively.

WORKS CITED

Alarcón, Norma, Caren Kaplan, and Minoo Moallem. "Introduction: Between Woman and Nation." *Between Woman and Nation: Nationalism, Transnational Feminisms, and the State*. Ed. Caren Kaplan, Norma Alarcón, and Minoo Moallem. Durham, NC: Duke University Press, 1999. 1–16. Print.

Altman, Janet Gurkin. *Epistolarity: Approaches to a Form*. Columbus: Ohio State University Press, 1982. Print.

Anderson, Benedict. *Imagined Communities*. New York: Verso, 1983. Print.

Brey, Joe. *The Epistolary Novel: Representations of Consciousness*. New York: Routledge, 2003. Print.

Brown, William Hill. *The Power of Sympathy*. New York: Penguin, 1996. Print.

Carretta, Vincent. "Questioning the Identity of Olaudah Equiano, or Gustavus Vassa, the African." *The Global Eighteenth Century*. Ed. Felicity A. Nussbaum. Baltimore: Johns Hopkins Press, 2003. 226–35. Print.

Davidson, Cathy N. *Revolution and the Word: The Rise of the Novel in America*. 1984. Expanded Edition. Oxford: Oxford University Press, 2004. Print.

Elliott, Emory. *The Cambridge Introduction to Early American Literature.* Cambridge: Cambridge University Press, 2002. Print.

Evans, Gareth. "Rakes, Coquettes and Republican Patriarchs: Class, Gender and Nation in Early American Sentimental Fiction." *Canadian Review of American Studies/Revue Canadienne d'Etudes Americaines* 25.3 (Fall 1995): 41–62. Print.

Felski, Rita. *Uses of Literature.* Malden, MA: Blackwell Publishing, 2008. Print.

Fluck, Winfried. "Reading for Recognition." *New Literary History* 44.1 (Winter 2013): 45–67. Web. 2 Feb. 2014.

Haraway, Donna. *Simians, Cyborgs, and Women: The Reinvention of Nature.* New York: Routledge, 1991. Print.

Harzewski, Stephanie. "Tradition and Displacement in the New Novel of Manners." *Chick Lit: The New Woman's Fiction.* Ed. Suzanne Ferriss and Mallory Young. London: Routledge, 2006. 29–46. Print.

Johnson, Joanna Webb. "Chick Lit. Jr.: More Than Glitz and Glamour for Teens and Tweens." *Chick Lit: The New Woman's Fiction.* Ed. Suzanne Ferriss and Mallory Young. London: Routledge, 2006. 141–57. Print.

Mallan, Kerry. "Space, Power and Knowledge: The Regulatory Fictions of Online Communities." *International Research in Children's Literature* 1.1 (July 2008): 66–81. Print.

Mulford, Carla. "Introduction." *The Power of Sympathy* by William Hill Brown and *The Coquette* by Hannah Webster Foster. New York: Penguin Books, 1996. ix–li. Print.

Myracle, Lauren. *ttyl.* New York: Amulet, 2004. Print.

– "ttyl." Laurenmyracle.com. http://www.laurenmyracle.com/books/#/ ttyl/. Web. 28 Nov. 2014.

Tyrrell, Ian. *Transnational Nation: United States History in Global Perspective since 1789.* Houndmills, UK: Palgrave Macmillan, 2007. Print.

Warner, William. *Licensing Entertainment: The Elevation of Novel Reading in Britain (1684–1750).* Berkeley: University of California Press, 1998. Print.

Wells, Juliette. "Mothers of Chick Lit? Women Writers, Reader, and Literary History." *Chick Lit: The New Woman's Fiction.* Ed. Suzanne Ferriss and Mallory Young. London: Routledge, 2006. 47–70. Print.

White, Ed. "Early American Nations as Imagined Communities." *American Quarterly* 56.1 (March 2004): 49–81. Print.

Winans, Robert B. "The Growth of a Novel-Reading Public in Late-Eighteenth-Century America." *Early American Literature* 9 (1975): 267–75. Print.

Cyber-Alternatives to Lived Identities

6 Digital Diasporic Experiences in Digital Queer Spaces

AHMET ATAY

As a diasporic individual, I entered into digital queer spaces because I was looking to find my own cultural identity and a community to which I could belong. I was searching for other diasporic and queer individuals with whom to talk to and identify. Perhaps I wanted to carve out a digital space where I could be and become without questioning and judging my every step. Like others who are physically and emotionally away from their ancestral homes but simultaneously belong to multiple homes (physically, emotionally, and digitally), I was searching for answers to help me understand the complexities of diasporic queer experiences. Digital spaces became "spaces within," and communities I belonged to on quick media became hybrid entities where digital interactions were woven into physical realities. Perhaps this is why my diasporic queer experiences appear as patched-together realities (the amalgam of online and offline realities), communities, experiences, and identities. Through both personal and scholarly lenses, this chapter seeks to explore the implications of digital queer intersections on diasporic bodies.

Introduction

Cyberspace allows individuals to create communities, express their ideas, perform activism, and, more importantly, create a sense of self (Bell; Jones, "Information"; Watson; Shaw). Steven Jones ("Information") argues that like each new technological innovation, computer-mediated technologies have been created to attempt to improve life. This is certainly true in the sense that computer-mediated technologies can assist individuals in creating transnational communities and global kinship. To a large degree, they affect how a person perceives others, relates to the world around him/her, and makes sense of his/her cultural identity. Furthermore, computer-mediated communication restructures some of our conventional understandings and identity performances, such as presentation of self, and the representation of sex and sexual acts. Quick media such as social network sites and smartphone apps offer alternative cultural spaces for different identities to emerge and to be performed.

Queer immigrants and diasporic queer bodies often come to social network sites to create alternative communities where they connect with others and escape from daily limitations and constraints because of their complex cultural identity, both sexual and ethnic. While these online communities often function as alternative communities, they are situated and also connected to other online and offline communities. Therefore, the nature of these communities is different than their offline counterparts because they are voluntary, organic, sometimes temporary and sometimes long-lived, and not bound by time and place. They also exceed national and linguistic boundaries. Therefore, these platforms are nationally and linguistically diverse entities that function as a common communication ground for sexually marked immigrant and diasporic bodies as well as for non-diasporic mainstream members. These themes resonate through my personal experiences with queer digital spaces as a diasporic actor.

In the Beginning

I don't remember how but I quickly became a member of two new communities in Hunt for Men and Gay Space after joining in 2008.[1] They remained distinct yet connected. I simultaneously existed within the two to cultivate a sense of belonging and also to communicate with others who could be defined as queer, queer immigrants, and queer diasporic bodies. While our bodies were racially, ethnically, and nationally marked differently, and while we used English as a common language to communicate physically, ideologically, and emotionally, we were dispersed. We were different but at the same time similar because

---

1 Here I use pseudonyms to refer to the original sites in order to protect the identities of their members.

we were all looking for companionship, friendship, and perhaps love and sex within these social network sites. While a commercial digital infrastructure created these sites for economic benefits, as queer bodies (both diasporic and non-diasporic), we served in them as marked labourers who were seeking digital communities away from our physical communities. Therefore, we created digital versions of ourselves to be communicated, displayed, and finally consumed by others who were benefiting from our digital presence on quick media and social network sites. I, for one, was looking for places to belong and others with whom to communicate. I was not alone in my journey to find other diasporic individuals within and outside of the United States.

Both Hunt for Men and Gay Space offered new possibilities that were not part of my experience with my former cyber communities. For example, the webcam-based chatroom function in these sites provided new communication modes to view others and also to be looked at. Gay Space provided access to other queer members from different geographical territories and European countries in particular. Therefore, these new capabilities offered alternative and diverse experiences which were not easily available through location-based chatrooms. These sites not only provided access to more diverse populations but also enabled the creation of more diverse queer cyber communities. Furthermore, they facilitated the emergence of diasporic co-cultures within the larger community as a result of the nature of the sites.

The meaning of queer started to blur as I interacted with other queer bodies whose experiences were shaped by different cultural practices and geographical and locational meanings that are attached to their bodies. Queer became a fluid notion whose meanings were embodied differently by diverse groups of people and their particular experiences.

Presentation of Diasporic Queer Self in Cyber Communities

Presenting oneself, as Ervin Goffman explains, offers meaning about us to others. Therefore, the ways in which we present ourselves to others in quick media project certain contextual and personal meanings about who we are. Goffman argues that "[i]n everyday life, of course, there is a clear understanding that first impressions are important" (11). The profiles we create in quick media and social network sites – the ways in which we present ourselves, our ethnic and racial backgrounds, national heritage, or different aspects of ourselves – communicate meanings about our identities. In return, our self-representations influence how we are perceived by others, the nature of our communication with them, and our unspoken negotiations in constructing a cyber-based queer community.

Creating a profile on a social network site is an important task because these profiles are representations of our identities. For example, the profile picture and personal information that I provide facilitate my entry to a cyber queer community. Therefore, this profile aims to present dimensions of my diasporic queer self, perhaps my "best self," to others. The main question when I began creating profiles in these new social network sites was how am I supposed to present myself as a diasporic queer body living in a small Midwestern college town to others (both diasporic and non-diasporic)? When the presentation of myself was completed, I was ready to be displayed and consumed by others. During this consumption process, I entered into a cyber community simply by allowing others to look at my profile. Similar to building offline communities and kinship, the act of looking functions as the first step in initiating communication with others. Looking as an act is typically followed by communicative practices. These communicative acts function as building blocks for online communities whose members are geographically dispersed. One's profile becomes an extension of one's identity to negotiate friendships and access to a larger community. My access to this online community began with one click.

Click ... click ... click. So many images of young and attractive men, some geographically close to my physical location and others miles away. While each click allowed us access to each other's lives, each click also functioned as a means of co-constructing a community, a group of queer bodies who were culturally and linguistically different.

Click ... click ... click. Each click helped me to access their profiles; each click helped me to open a cyber window to facilitate communication between us. "Hi ... where are you? Would you like to chat?" As replies appeared in separate windows, I connected with a person who was miles away. While I watched others have conversations in the main chatroom, I conversed with different men in private rooms. I looked at their pictures while they consumed mine. I looked at their bodies that were on display – shaped differently, pictured differently, displayed differently. Some pictures were in public while others were shown in private. Pictures of different sizes and shapes of penises were locked in these private boxes. While I co-constructed a cyber community through engaging in different online communicative acts, I had access to their lives and their bodies through images and words that appeared on my screen. Co-creating a cyber community composed of geographically and linguistically diverse individuals was radically different from entering into a physical community that was bounded by space, location, time, and language. Therefore, differences in community building should be acknowledged in order to theorize the notion of online communities, and particularly the ways in which diasporic queer individuals negotiate their memberships in these communities.

Click … click … click. Community was in the making. Each conversation screen that appeared and each profile that I looked at connected me to others, other queers who were immigrants, diasporic bodies, racially and ethnically marked selves, and bodies who enjoyed privilege because of their skin colour and income. One click helped me to construct a series of multinational and multicultural global communities.

First, I resided in the Illinois room. My first cyber-home. Shortly after, I began talking with people in other rooms. It took some time to identify the regulars as the members randomly came and left. Only a few of them stayed long enough to participate in the construction of my new online communities. Since quick media forms often allow individuals to appear and quickly disappear in different online platforms, this particular act creates two types of members: ones who briefly use these platforms and quickly disappear and the others who stay longer and construct long-term relationships and more stable communities.

Due to geographical proximity, I talked to several diasporic people in the St Louis area. They were mainly immigrants who came to the United States during and after the war in Bosnia in the 1990s. Some of them were not out to their families and therefore only existed as queer bodies in cyberspace. Their lived experiences and concerns were different from their non-diasporic counterparts'. While most of the US members were concerned about their self-representation, their physique, and the ways in which they were perceived by other queer members on these social network sites, some of these Bosnian immigrants, who were part of a diasporic community in St Louis, were worried about fitting in, finding a community. Their concerns were mostly about negotiating between the diasporic communities that they were physically contained within, the mainstream white American ideals that surrounded them, and the queer communities within their geographical location. They were trying to make sense of their identities by using these social network sites. At the same time, they were seeking others who were going through similar experiences, or who were challenged and affected by queer communities around them. Therefore, the ways in which they contributed to the community-building process and kinship was different. They stayed longer. We stayed longer. We had conversations about queer identities, fitting in, social and cultural expectations, and fears that guided and influenced our realities, both online and offline.

Now I was part of a marginalized cyber community within a cyber queer community. While our experiences often mirrored each other's, our conversations helped us to make sense of being a diasporic queer body within a larger mainstream queer culture, whose rules and practices often can be oppressive towards minorities and immigrants. We never met outside of our immediate cyber community; however, we bonded over issues and experiences we have had. This community lasted a long time because of the need to belong to a place, to a group, and to a community.

This community dissolved as quickly as it was built as the site recreated and reshaped itself. Cyber communities are dependent upon the capabilities of quick media and social network sites. They can easily disappear as technologies and platforms evolve and change.

Finding Home

Notions of home and belonging carry different meanings for diasporic individuals who are dispelled from or left their home countries compared to their non-diasporic counterparts (Iyer; Hall; Sarup; Trinh; Rushdie). These meanings are contextually bounded and profoundly influenced by the individual's gender, sexuality, nationality, ethnicity, and socio-economic background. Because of their hybrid identities, diasporic individuals might occupy more than one home and belong to more than one nation-state, and they create different ways of being or feeling at home based on their experiences and backgrounds. For example, Trinh writes, "For a number of writers in exile, the true home is to be found not in houses, but in writing" (16). Similarly, Hall, Rushdie, and Iyer argue that diasporic individuals use writing, ideas, and languages to feel at home or find ways of belonging to a community, a culture, or a nation. In this chapter, I argue that diasporic individuals, and diasporic queer bodies specifically, use quick media platforms and Internet technologies to represent identities, generate a sense of belonging, and create online communities based on the idea of forming digital kinship. Since these new and consistently emergent communities are often created through quick media outlets and cyberspace platforms, they exceed the borders of nation-states and problematize the notion of proximity in relation to community by collapsing the disparities between time and geographical locations.

These digital queer communities are built on the idea of being patched together. They are digitally infused alliances because they shelter a mixture of individuals who speak different languages and physically reside in different online and offline locations. Therefore, diasporic queer individuals build a digital kinship despite being in different places and speaking disparate languages. Although they may speak different languages, they use English in the formation of these imagined communities to connect them together and help them to create a community.

Digital Queer Communities

Even though geographical proximity is still a determining factor in community building in some queer quick media outlets and social network sites, certain platforms allow members from different parts of the world to share a communal digital space and create a global community. In this way, there are differences in platforms, their nature, and technological capabilities. These technological aspects also play a paramount role in the creation and sustainability of a community. While some location-bound quick media platforms allow their members to communicate with others within their geographical location, these members might never form a community despite using these platforms to communicate. Because quick media platforms, such as downloadable apps, facilitate quick formations of online communities, these communities can also quickly dissolve when the members are geographically dislocated. On the other hand, other platforms might allow members, despite their geographical barriers, to form and sustain a digital community. The latter platforms enable diasporic queer individuals to communicate with each other despite their geographical differences. Furthermore, these platforms also allow these individuals to communicate and build communities with queer individuals from their host nation as well as their countries of origin. Engagement with these types of communities and platforms were my point of entry to this topic.

My introduction to queer social network sites and my involvement within queer cyber communities began with GayCommunity, a quick media platform for gay men to meet with others within their geographical location as well as men in other areas both nationally and globally.

As a graduate student, I spent most of my time in small Midwestern college towns. Even though I belonged to different social communities and counted mostly international and American students and faculty as my friends, until my coming out, I lacked the support of queer communities and friendships with other gay men. Therefore, GayCommunity became a safe haven where I conversed with others who could know me as queer. GayCommunity was a geographically bound site at the time. There were chatrooms for each state, further divided into cities with their own chatrooms. Typically, members had conversations with other members who were physically proximate. Even though most of my conversations were limited to simple chats, some of them were directed towards more sexual content. At first, I was only a browser, uncommitted and without a sense of belonging. For me, members of these sites only existed as cyber entities, bodies that appeared only in cyberspace. While I browsed through Web profiles that included pictures and detailed descriptions, mine lacked obvious identifiers, such as a photograph. The lack of a picture in my profile signified my invisibility within the queer community. My picture-less profile represented my state of being. As I was in the closet, so was my picture. I only existed within the boundaries of GayCommunity, as if I were just an illusion. I was a faceless entity.

In time, I learned the value of identifiers, the symbols, images, and writing that members use to represent themselves. Descriptive language and facts about a member, such as height, weight, location, and relationship status, mattered in these cyber platforms. The lack of such information and more importantly a picture often resulted in denial of access to a larger community and fewer possibilities of meeting with others, particularly potential relationship partners. As I moved from one geographical location to another as a graduate student, my physical offline communities and friendship circles changed, but the online spaces also transformed. When I moved from Iowa to Illinois, I was not only uprooted from a physical location that I called home for two years during my graduate education, I was also digitally dislocated.

Defining Community

The notion of community functions as "a central construct" in envisioning and defining the ways in which humans organize their lives, create a shared reality, and make sense of their communal experiences (Mitra 55). Like Mitra, Jones also argues that community is a culturally constructed category. Hence, as Jones writes, "[c]reating and maintaining community has traditionally been valued as a commendable goal" ("Information" 23). Communication plays a vital role in creating, recreating, and maintaining the idea of community (both online and offline) (Watson 104).

Communities may be understood in many ways, but three notions of community stand out: (1) geographical location, (2) sense of belonging, and (3) imagined communities. According to Doheny-Farina, "A community is bound by place, which always includes complex social and environmental necessities. It is not something you can easily join. You can't subscribe to a community as you subscribe to a discussion group on the net. It must be lived. It is entwined, contradictory, and involves all our senses" (37). Even though in the traditional sense a community is described in relation to a shared geographical location, often it is described as "belonging" to a larger social and cultural group. According to Bell and Valentine,

> "Community." It's a word we all use, in many different ways, to talk about … what? About belonging and exclusion, about "us" and "them." It's a common-sense thing, used in daily discussions, in countless associations, from "care in community" to the Community Hall; from "common spirit" to the "business community" … Many of us would lay claim to belonging to at least one community, whether it is the "lesbian and gay community" or just the "local community" where we live … [T]he term community is not only descriptive, but also normative and ideological: it carries a lot of baggage with it. (93)

Hence, the idea of community is constructed around the notion of "belonging." In his widely acknowledged argument, Anderson described communities as being imagined. He notes, "Communities are to be distinguished, not by their falsity/genuineness, but by the style in which they are imagined" (6). While we might share a geographical location, our connection with other members is only imagined. As long as we have a commitment to this idea, we function as a member of a community (Jones, "Information" 4). Even though there are similarities between offline and online communities, they are also distinct. Sharing a common geographical place is no longer a required characteristic for defining a community. Though individual members of these sites and platforms might also "belong" to a community based on residing in a shared geographical location, they are still often committed to their online communities as well.

These communities are organic and ever changing, but there are similarities and differences among them. While some of them are long-term communities with regular members, some are temporal in their nature. Temporality or regularity of the communities is often shaped by the nature of the quick media platform but also the nature and composition of the group members. For example, a group of close friends might utilize Skype to communicate regularly. In this example, the community that is formed is a long-term one. On the other hand, other communities formed in social network sites (queer or not) might be temporal because the members of the community can easily depart (both geographically and digitally) and move into a different geographical location or digital site. Unless the members are committed to these online communities and they successfully build long-term relationships, their presence will be temporally bounded. Furthermore, members of online communities might belong to multiple communities simultaneously and they might voluntarily manoeuvre around them while cultivating a sense of belonging in each community. Potentially, the kinship that is formed among the members of these online communities might exist within and outside of the boundaries of quick media and social network sites. Just as some of these communities could be transformed into offline relationships, some might have started as offline communities and moved into online domains. Despite these differences in the formations of communities, quick media and social network sites function as a vehicle or platform to help queer bodies, both diasporic and non-diasporic, to create social and cultural kinship and a sense of belonging to a larger cultural collective outside of their immediate offline communities.

The community that I built in GayCommunity was left behind the moment I crossed the invisible state border between Iowa and Illinois. I do remember that while I was settling down in my new environment, I often connected to the former location-based chatrooms, which I referred to as my cyber community and cyber home, for comfort and familiarity. Feeling isolated, I looked for a sense of belonging in quick media and online platforms. The screen names and digitalized pictures of members indicated a familiar space through which I was able to find comfort and connect with others. The companionship of cyber friends and communities that I once belonged to provided a point of connection. As time passed, our online community changed and transformed because of movements of bodies. Some of the regular members moved away physically (graduated or took a job in another location) and digitally (immigrated to other social network sites or quick media platforms). Therefore, every physical move that I made meant the disappearance of a location-based cyber community and its members, despite us having never physically met.

With time and geographical disconnection, the quality of my relationship with location-bound cyber communities decreased as I created new online and offline communities that resulted from changes in members' lived experiences (both online and offline), emergence into new digital communities, and lack of commitment in digital kinship. I believe that unlike our offline communities, which may require shared common history and location, prior physical contact, and commitment, some online communities can easily dissolve from lack of physical contact and commitment in kinship. As my loyalty to my previous online community lessened, the online friends I once had became a distant memory. Considering that our cyber environment and our online presence within social network sites and quick media platforms has shifted, changed, or completely disappeared, my experiences of this digital place and our meaningful interactions can be remembered but not physically recovered. Since our sense of community and interactions were limited only to cyber environments and these interactions were not carried out in offline venues, most if not all of these relationships existed in a temporal frame and were composed of digital bites. As these social network sites and quick media platforms evolve and change, digital communities that are situated in these venues also shift based on change and technological developments.

In September 2003, I joined a new digital community. As a diasporic queer man who wanted to come out, I used this community as a safe haven alongside the physical peers who supported my coming out process. During this time period, I held long-term conversations with others who were also residing in the same cyber room. Like the previous ones, this chatroom was geographically bounded and composed of members living in my town or surrounding areas. I was fearful of seeing them outside of these online platforms. The idea of mixing my online realities with offline ones and crossing the boundaries between digital and physical spaces was a scary thought. Even though there were some diasporic individuals among the members, most were non-diasporic bodies. As time progressed, some of these offline interactions happened as our paths crossed at local bars, grocery stores, university-related functions, and sometimes unexpectedly at random locations.

This time, I sustained a long-term location-based online community because of the duration of my doctoral work. As members came and went, the nature and dynamics of my online community changed with these arrivals and departures. Unlike my last location, my new institution had a large number of international students, at both the graduate and undergraduate levels. As a result, some of these students became regular members of the cyber communities to which I belonged. Like me, these members were using social network sites, such as GayCommunity, and our chatroom in particular to negotiate their identities and look for friends, sexual partners, or long-term love interests. They were also using this space to express aspects of their identities as they tried to make sense of their sexualities. Their presence on this site was limited to chatroom conversations; however, they were part of a larger location-bound collective as the members were building digital kinships. The site became a venue for support, friendship, and sexual exploration.

Virtual, Local, Global

During the fall of 2008, I discovered Hunt for Men and Gay Space, two new social network sites that are particularly geared towards queer members. These platforms go beyond the idea of geographical location-based chatrooms and create language-based or regional chatrooms where members can easily converse with others outside of their immediate location. Most of these sites also allow members from different parts of the world to communicate with one another though chatroom conversations as well as webcam communication. Hence, the global reach of these sites helps their members to construct different global queer communities that are geographically, nationally, and linguistically dispersed.

Location-bound social network sites like GayCommunity still exist; however, there is an increasing presence in global chatrooms where members from different parts of the world can communicate. On the other hand, the presence of location-bound platforms is more visible in smartphone-based quick media apps, such as Hornet and Grindr. Even though members of queer communities often use these platforms to meet new people or find sexual partners, these communities differ widely from communities on social network sites. Despite the fact that close proximity is an important factor in the community-building process, the goal of these quick media forms, or their capabilities, means that communities that might form would not necessarily have the characteristics of traditional communities. Communication between members revolves around private messaging, and there is a lack of communal discussion platforms where different members can converse. Because of these characteristics, most of these apps do not effectively participate in community building but are instead used as quick communication devices. As I travelled, figuratively and literally, through the process of both coming out and engaging in diaspora, I began to explore new sites as spaces for virtual community and kinship.

Finding New Digital Homes

As an international social network site, Hunt for Men offers Web profiles, instant messenger, and webcam-based communication forms to its users. The site currently has over 1.5 million profiles worldwide, over 1 million of whom are active, meaning the members are actively participating in chatrooms and regularly updating their profiles. Chatrooms at Hunt for Men are divided into regions, languages, countries, or interests. Each main country is also divided into regions, municipalities, or cities. For example, the United States is divided into smaller regions, and some states are represented individually. Each represented state is broken down further into towns. Members can also search for others in their immediate areas by performing searches based on zip codes or town names. Therefore, even though they might not be chatting with others local to them collectively in a chatroom, they can still communicate with members through instant messages and webcams. However, this site still mostly functions as a location-based social network site. Hence, communities that are formed likely still share similar locations.

Recently, the website decided to refigure their chatroom layout and removed smaller and relatively uninhabited chatrooms to make room for language-based rooms. For example, the members of the English room are composed of primarily British, Canadian, Australian, and US members. However, there are also members residing in these rooms who come from countries whose official language may not be English. Some of the chatrooms are also organized around a particular interest. These rooms often contain members who come from different geographical locations but share similar interests, both sexual and non-sexual.

Although during the earlier days the quick media platforms and social network sites were limited to only picture- and text-based communication, in the last couple of years, webcam-based communication capabilities have also been added to the site. Therefore, communication among members has become complex and interactions between members have begun to take place in multiple forms because webcam-based interactions now allow participants to communicate without using written communication. However, members can still use more traditional forms and webcam-based communication simultaneously. These new technological advancements influence the nature of communities by making multiform communication available. The level of intimacy and intensity of communication might increase among members because mediated face-to-face communication often allows members to see each other and talk with each other without using discussion boards. These forces influence the ways in which digital communities form and their members interact.

In order to register for Hunt for Men, members only need a valid zip code. When one becomes a member, he can communicate not only with people in his immediate city but also with other members who reside within the state or in different regions. In order to communicate with other members, one needs to enter into rooms that are dedicated to different states, towns, or regions. Each member needs to create a profile and include a picture; however, using a webcam is optional. While users identify themselves by their zip code, their communication with other members is not limited to geographical areas in that they have their choice of chatroom. Paying members of the site are granted limitless messaging abilities and permission to reside in multiple rooms simultaneously, while non-paying members are provided only limited access and capabilities. Due to the nature of the site, the users are mostly located in the United States, and only some of them are diasporic in nature.

Gay Space is a European-based social network site. Compared to GayCommunity and Hunt for Men, Gay Space has more international visibility and functions more like a social network site rather than a queer lifestyle online magazine. The site is also more popular in the United Kingdom, Ireland, continental Europe, Australia, and South Africa than in North America. Like other sites, Gay Space provides a space for queer bodies to communicate and create their own cyber spaces. However, the sites are tailored for different users even though on the surface they might appear similar. While some of these sites are funded by advertisers (such as sex shops, sex toy companies, and porn companies), some of them have membership fees. Noticeably, Gay Space does not come across as a commercially driven network, even though it depends on membership fees and other form of external funding. In addition, Gay Space uses different features, such as a radio channel that plays the most popular North American and European songs. Similar to other sites, Gay Space also has a common discussion board; however, most of the communication between members happens through private messages. In addition, even though members of the site are still bounded by geography, they can easily communicate with members from other locations.

One of the most distinct differences between Gay Space and Hunt for Men is that Gay Space offers services in different languages (English, German, French, Italian, Spanish, Dutch, Portuguese, and Japanese). This feature clearly illustrates the site's international reach. A valid user name and password are enough to become a member. Similar to GayCommunity or Hunt for Men, one has to be over 18 years old to be legally accepted as an active member. To register, one has to choose a country of residence. As with other social network sites, Gay Space requires detailed background information from users, such as location, whom they are interested in meeting (single man, single woman, gay male couple, gay female couple, bi couple, or group), and other information ranging from birthday, body type, profession, drug and alcohol consumption, and smoking status to penis size, ethnic origin, and several other pieces of information. The intention behind these questions is to capture and provide more realistic self-portrayals of the members and their identities.

The discovery of these social network sites enabled me to belong to multiple cyber-based communities at the same time. Unlike GayCommunity, Gay Space and Hunt for Men allowed me to connect with people from different parts of the world with different cultural backgrounds. For the first time, I was able to communicate with other queer immigrants and diasporic bodies who reside outside of the United States, which allowed me to understand experiences of diasporic and non-diasporic queer individuals in different cultural contexts.

Home and Diasporic Queer Communities

Since "home" is typically considered to be the place where one belongs, I see home and belonging as two interrelated concepts. This discussion suggests that one can occupy more than one place, both geographically and psychologically; therefore, one can belong to more than one location, both geographically and digitally.

Diasporic individuals might occupy more than one home and might belong to more than one nation-state, and they create different ways of being or feeling at home. In this context, new media technologies in general and quick media forms and social network sites in particular facilitate a sense of belonging by allowing multiple alternative and cyber places and homes to simultaneously coexist by enabling diasporic and non-diasporic individuals to create communities outside of their geographical location. Moreover, members of these sites can simultaneously belong to multiple cyber communities and manoeuvre around them as they establish a sense of belonging and seek home and kinship. Even though some of these attachments are temporal because cyber communities can be established or dissolved quickly, some of them are long term, and diasporic individuals often enjoy the idea of belonging to a community whose members share similar lived experiences and cultural backgrounds. My personal experiences echoed these notions of digital kinship.

The notion of queer only existed for me within the context of US culture, and my involvement in queer communities was limited to my presence in queer quick media and social network sites and infrequent appearances at local gay bars or social events. Access to different global queer communities meant a lot. I was able to communicate with other queer individuals whose experiences were affected by their location, diasporic or immigrant status, and the nature of the queer community in their host nations.

At first, I was confused, perhaps a bit afraid of talking to queer individuals who shared a similar cultural background and language. I only existed as a queer body within the US context and through the English language. Previously, when I interacted with other queer bodies on these social network sites (international, diasporic, immigrant, or not), I only communicated with those who came from cultural backgrounds different from my own. One click would be changing a lot, I thought.

Three clicks and I was in three different chatrooms: a general chatroom from Turkey, one from Cyprus, and I could not resist the temptation of entering the London chatroom. I was not looking for anything in particular. All I wanted to do was look at the Web profiles of members who spoke the same language or shared common cultural experiences with me. Was I looking for a community, a community away from others, physically and culturally similar yet different? Even though their physical locations were thousands of miles away, we met at a cyber location.

I do remember my first reaction to this cyber experience. I was confused, to say the least. I was worried about the geographical disparity between myself and other members. I lurked around and lingered for weeks, but it took me some time to initiate conversations. To my surprise, my presence at these sites did not last long. As a diasporic queer body away from my home country, I was not able to relate to the other members. Their experiences and mine were different, as if we were speaking different languages, though obviously we were narrating different experiences and stories that were not even slightly intersecting. Geographical difference mattered after all. In this case, my attempt to create a community away from my own, culturally similar yet geographically different, failed. After all, I was away from the motherland, and being queer only made sense within the US culture; as a diasporic queer individual my experiences of queerness has always been influenced by my diasporic status.

Living away from the homeland changes and shifts the identities of diasporic bodies in a particular way. Not only their identities but also the meanings that they attach to their homelands shift. Even their home cultures become different. While they move away from their past cultural practices when they are in different locations, they come closer to the culture of host nations. However, this blending often results in confusion and alienation from both cultures. I was at the intersection of those experiences. I didn't entirely belong to the cyber communities that I encountered in these social network sites, which exceeded the physical boundaries of my new cultural location. Dis-identification with the cultural practices and language facilitated communication between us. Members of the site from the homeland, who were distinct yet similar, led me to grow apart from the communities in which I sought shelter and looked for cultural connection. Instead I turned towards communicating with others whose experiences, like mine, are in-between, complex, and always negotiated between queer, diaspora, and language contexts.

When I left Illinois, I also left behind a community whose members were composed of Illinois natives and others from different parts of the world. Though we were free wanderers – partly diasporic, partly cosmopolitan, and partly native – we were able to cultivate a community away from our own geographical location. This time we were able to create social bonds because of differences: cultural, linguistic, political, and legal.

I have been living away from Illinois for a while now. The communities I constructed both offline and online have become only a distant memory.

While the members of those communities have once again dispersed and moved away, they also created new communities bounded in location in cyberspace. The communities we left behind were temporal, formed out of necessity, and due to geographical proximity. When we move, as diasporic bodies, between cultures, languages, and locations, we create new communities and identities that are co-constructed by different members. Like our identities, our communities and the stories that we tell about them are collages of memories, pictures, texts, experiences, and fragmented histories.

Returning to Cyber Homes

Since I have been living in new digital communities, I never returned to the location-bound digital communities to which I once belonged. Now, while I converse with others on quick media platforms and social network sites, I reside in non-location-based global chatrooms and social networks. Every once in a while I use location-based smartphone apps to communicate with other queer individuals around me, but I do not consider our communication as kinship or our presence in these platforms as communities. Our interactions are limited, fragmented, and temporal. After four years of absence, I recently checked out Gay-Community, partly because I was doing research for this chapter and partly because of an urge to revisit the past. When I started to type my screen name, I hesitated. I was not even sure if that was the correct screen name. Remembering the password was not easy either. After a number of failed attempts, I managed to log in. To my surprise, nothing looked the same. The colour scheme of the site, the main page, and the functions that once were available are now gone or functioning differently. I did not even recognize my own profile. Other than my picture from four years ago, everything looked and felt strange, alien, and very unlike what I remembered. My search for my former communities also failed. Even though I was able to enter the chatrooms that once felt like home, they were different than before. None of the members looked familiar. I felt like a stranger.

Because I considered these chatrooms homes away from home that existed only in cyberspace, I was stricken by a sense of loss, a feeling as though I had lost something very dear to me. Even though I have not been a regular member of the site for the last four years, returning to my cyber "home" meant losing something that once felt familiar, comfortable, and comforting.

Conclusion

Although hybrid positions might provide an opportunity for diasporic individuals to move between cultures, to reside simultaneously in multiple geographical and psychological locations, and to occupy different standpoints that are often enabled because of the presence of quick media technologies, such individuals still remain marginalized within their communities. In the case of diasporic queer bodies, this could extend to marginalization within their own diasporic communities as well as mainstream queer communities. Hence, they negotiate the notion of digital kinship differently because of their lived experiences and patched-together identities. I argue that digital platforms and social network sites enable the facilitation and representation of diasporic queer identities. Since diasporic queer bodies experience constant flux, the state of hybridity can be considered a fluctuating state of being that allows contestation, negotiation, and (re)creation of diasporic cultural identities. Consequently, through these digital and liminal spaces and hybrid states of being, diasporic queer bodies carve out physical and psychological locations in which to exist within their host and diasporic cultures simultaneously and create globalized digital queer communities where they can belong and express aspects of their identities and digital kinship.

WORKS CITED

Anderson, Benedict. *Imagined Communities: Reflections on the Origin and Spread of Nationalism*. London: Verso, 1983. Print.

Bell, David. *An Introduction to Cybercultures*. London: Routledge, 2001. Print.

Bell, David, and Gill Valentine. *Consuming Geographies: We Are Where We Eat*. London: Routledge, 1997. Print.

Doheny-Farina, Stephen. *The Wired Neighborhood*. New Haven, CT: Yale University Press, 1996. Print.

Goffman, Erwin. *The Presentation of Self in Everyday Life*. New York: Anchor, 1959. Print.

Hall, Stuart. "The Whites of Their Eyes: Racist Ideologies and the Media." *Gender, Race and Class in Media*. Ed. Gail Dines and Jean M. Humez. Thousand Oaks, CA: Sage, 2010. 18–22. Print.

Iyer, Pico. "Living in the Transit Lounge." *Unrooted Childhoods: Memories of Growing Up Global*. Ed. Faith Eidse and Nina Sichel. Yarmouth, ME: Intercultural Press, 2004. 9–23. Print.

Jones, Steven G. "Information, Internet, and Community: Notes towards an Understanding of Community in the Information Age." *Cybersociety 2.0: Revisiting Computer-Mediated Communication and Community.* Ed. Steven G. Jones. Thousand Oaks, CA: Sage, 1998. 1–34. Print.

– "Studying the Net: Intricacies and Issues." *Doing Internet Research: Critical Issues and Methods for Examining the Net.* Ed. Steve Jones. Thousand Oaks, CA: Sage, 1999. 1–27. Print.

Mitra, Ananda. "Virtual Commonality: Looking for India on the Internet." *Virtual Culture: Identity and Communication in Cybersociety.* Ed. Steve G. Jones. Thousand Oaks, CA: Sage, 1997. 55–79. Print.

Rushdie, Salman. *Imaginary Homelands: Essays and Criticism 1981–1991.* London: Penguin Books, 1992. Print.

Sarup, Madan. *Identity, Culture and the Postmodern World.* Athens: University of Georgia Press, 1996. Print.

Shaw, David F. "Gay Men and Computer Communication: A Discourse of Sex and Identity CyberSpace." *Virtual Culture: Identity and Communication in Cybersociety.* Ed. Steve G. Jones. Thousand Oaks, CA: Sage, 1997. 133–46. Print.

Trinh, T. Minh-ha. "Other than Myself/My Other Self." *Travellers' Tales: Narratives of Home and Displacement.* Ed. Jon Bird et al. London: Routledge, 1994. 9–26. Print.

Watson, Nessim. "Why We Argue about Virtual Community: A Case Study of Phish.Net Fan Community." *Virtual Culture: Identity and Communication in Cybersociety.* Ed. Steve G. Jones. Thousand Oaks, CA: Sage, 1997. 102–33. Print.

7 Claiming Ourselves as "Korean": Accounting for Adoptees within the Korean Diaspora in the United States

KIMBERLY McKEE

Technological communications across space, place, and time continue to reinvent everyday understandings of community. Members of a diaspora are no longer bounded by physical geography. Rather, these various constituencies can forge bonds and unearth new connections with one another in previously unknown ways. In examining how the Internet provides a venue for reincorporating "lost" voices within ethnic communities, this chapter explores how adult Korean adoptees raised by predominantly white families locate themselves within the Korean diaspora in the United States (Turkle; Wilson and Peterson). An estimated 130,000 Korean children entered the United States following the end of the Korean War (1950–3). These immigrants join Korean military brides as the two largest groups of the Korean diaspora in the United States (Yuh). Nevertheless, even though adoptees represent 1 in 10 Korean Americans, it was not until the last 20 years that the adoptee experience was inserted within broader understandings of "Korean American" or "Asian American."

Recognizing that cyberspace is a place where race remains salient, I am concerned with how adoptees negotiate their identities as culturally white, racially Asian individuals vis-à-vis their assertion of a Korean/Asian American identity. As digital communication continually responds to new technologies and social networking sites, the ease with which information flows between individuals presents additional avenues for deterritorialized communities to become reconstituted in cyberspace. The Internet offers adoptees an avenue to investigate their intersectional identities without the constraints of geographic location hampering mobility and offline connections. Not only have adult adoptee organizations coalesced on Facebook via the "Groups" and "Pages" features, but adult adoptees worldwide maintain deterritorialized relationships as Facebook "friends." In November 2014, domestic and international adoptees came together on Twitter to assert their voices during Adoption Awareness Month with the #FlipTheScript campaign. Originating with the adoptee-centric, independent writing collaborative Lost Daughters, the hashtag seeks to re-centre the voices of adoptees as important to adoption. More broadly, in the case of Korean American adoptees, the online publications of *KoreAm* and *Hyphen Magazine* alongside the popularity of Asian American–focused Web 2.0 media (e.g., Angry Asian Man, 8Asians, 18MillionRising) provide new chances for them to connect with Korean and Asian American communities.

To account for how cyberspace provides adoptees multiple lenses to assert and cultivate their identities as Korean or Asian Americans, this chapter analyses adoptees' participation on the ethnic website IAmKoreanAmerican.com. The site connects Korean Americans from across the world and highlights the population's diversity. Initially designed only to allow individuals to submit profiles that articulated their sense of Korean American identity, IAmKoreanAmerican.com expanded to include musicians, entrepreneurs, and projects/causes. This site reflects wider shifts in society concerning information dissemination, as seen in the growing popularity of Twitter, Tumblr, and Instagram. Web 2.0 media has increased individuals' ability to communicate and sustain community across space and time. Recognizing the public nature of the Internet, the site's participants have a clear sense that anonymity is absent as their information and pictures will be available for public consumption.[1] As part of their autobiographies, some participants feature links to their own personal/professional websites, YouTube channels, and Twitter feeds. Individuals featured on the site represent a new form of online participant – one keenly aware that anonymity, whether real or perceived, no longer exists in the second decade of the twenty-first century. While the interaction

---

1 This is evident in the site's privacy policy. When submitting their profiles, individuals must provide their full name, biography, photo, and location. Each profile's Web address also contains their full first and last name. Nonetheless, in order to maintain a semblance of anonymity for those featured on the site, I only identify adoptees by their first name and last initial, even as the site provides viewers each participant's first and last names, photo, and location.

between individuals is passive in that profiles are viewed and any further interaction (i.e., engaging with the user's YouTube channel or Twitter account) is possible but not mandatory, individuals who visit the site find themselves asserting a particular type of relationship bounded by ethnic solidarity.

IAmKoreanAmerican.com offers adoptees the opportunity to intervene in conversations that historically overlook their presence in discussions of what it means to be "Korean American." The site is a rich source of material to understand how adoptees reconcile tensions concerning their legitimacy and authenticity as "Korean" because of their transracial upbringing, in light of the fact that an estimated 75 percent of adoptees entered white families. Individuals actively participate in the assertion of a Korean American identity by submitting their profiles for inclusion. Instead of adoptees using the anonymity provided by the Internet, their engagement with the site provides new opportunities for inclusion as Korean, and even Asian, Americans (Danet). These individuals cultivate a new online community by creating an alternate avenue to assert one's claim to "Koreanness" outside traditional offline means of joining organizations or clubs based on ethnicity (Baym).

This chapter draws from 56 adoptee profiles out of 310 Korean American profiles posted on the site from 6 October 2009 to 3 May 2011.[2] Adoptees range in age from 19 to 58 years old at the time of their submission and reside in 16 states across the United States as well as in Germany and South Korea. Featured adoptees include well-known LGBTQ activists, a professional Olympic skier, comedians, and others actively involved in the adult Korean adoptee community. As part of their profiles, adoptees submit brief biographies, which are listed under the heading "Own Words." In my analysis of their autobiographies, I discovered common keywords/themes, including kimchi; internalized self-loathing for being ethnically Korean as youth; return to Korea; birth family searches; reunions with birth families; telling their adoption stories; involvement within the Korean community in Korea and in the United States; and activism. Similar to the ways in which a shared voice may emerge from individual blogs, the common keywords/themes found the "Own Words" section produce commonalties also found within the blogosphere (Mitra). While the 56 profiles explicitly noted the participants' "adoptee status," the level at which each individual

---

2 Two additional profiles featured adoptees under the age of 18, which will not be used in this study.

discussed the intersection of adoption and their assertion of a Korean (American) identity varied. For example, whereas some adoptees may discuss their return to South Korea, not all adoptees travel to South Korea nor desire to do so. These multiple voices reveal how the concept of "adoptee" is not monolithic. Their "Own Words" narratives reflect the myriad of adoptee experiences.

This case study provides the opportunity to better understand how adoptees self-identify and label themselves as adopted and Korean. While anthologies, memoirs, and documentaries are other venues where the public witnesses adoptees' negotiation of identity, I suggest that publishers and editors may mediate these forms of written text. Further, unlike adoptee-centric blogs, profiles on IAmKoreanAmerican.com are produced in a venue that positions adoptees as part of the Korean American community versus the adoptee community. While these two communities are not mutually exclusive, this analysis is invested in how transracial adoptees explore and assert ethnic identity. The site serves as a vehicle to highlight how the Korean American community is indelibly marked by the presence of adoptees as 1 in 10 Korean Americans are adopted persons. The inclusion of adoptees' voices allows them to forge kinship with a community that they may have lacked direct access to as they grew up in predominantly white communities. To disentangle adoption from ethnic identity would overlook how they are imbricated in the lives of adopted Korean Americans. IAmKoreanAmerican.com also offers adoptees the possibility to explore the faces of Korean America from the safety of their screens. While seeking kinship with other adoptees or Korean Americans more generally, adoptees can simultaneously find intimacy in the discovery of others like them and maintain distance.

Negotiating "Koreanness" Online: Inserting Adoption into Asian America

Adoptees' inclusion on the site highlights the heterogeneity within the community, while also recognizing the tie that binds these particular individuals – adoption from South Korea. Each profile deploys what Eleana J. Kim calls contingent essentialism, reflecting how "adoptee identity is at once essentialized as something natural and also construed as something cultural or socially constructed" ("Remembering Loss" 86). To be adopted means to be part of an imagined community, a wider family of Korean adoptees. Capturing this sentiment, Belgian adoptee Sunny Jo uses the term "KAD nation" to describe how adoptees create their own adoptee-specific cultural identity as they explore what it means to be Korean (Jo). A distinct subgroup of the Korean diaspora, adoptees form a non-normative family that exists not by bio-genetics but rather by their legal position as adoptees. Adoptees thus are linked within a postmodern kinship structure that recognizes how the community "takes on biological associations despite the inherently nonbiogenetic basis of adoption" (Eleana J. Kim, "Remembering Loss" 86). Forming the link that ties this particular population to the Korean American community, adoption cannot be viewed as an isolated act without repercussions. Rather, adoption underpins adoptees' kinship with one another and the creation of a KAD nation and, subsequently, a unique subgroup within the Korean diaspora.

The declaration of the phrase "I am Korean American" allows adoptees, a previously unincorporated population within both the Korean American community and the Korean diaspora, the ability to claim a right to their ethnic identity. This performative act reflects the power of language "to assert a true and inclusive universality of persons" (Butler, *Gender Trouble* 120). Adoptees' utterance of the phrase produces and harnesses the power of inclusion and exclusion concerning who is "Korean." To this end, I draw on Judith Butler, who writes, "Implicated in a network of authorization and punishment, performatives ... not only perform an action, but confer a binding power on the action performed" (*Bodies that Matter* 225). Reclaiming their identities as "Korean" and "American," adoptees destabilize the inherent meaning of "Koreanness," which is predicated upon biological/social parents of Korean descent. The historically static identity category is disrupted to account for non-normative ethnic Korean bodies. At the same time, adoptees' performative act markedly differs from another type of virtual racial performativity, identity tourism, which involves online users "adopting personae other than their own online ... [using] race and gender as amusing prostheses to be donned and shed without 'real life' consequences" (Nakamura 13–14).

The simple declaration "I am Korean American" is also a rebuttal to the South Korean government's decades-long inaction towards its adoption program and growing adult adoptee population. Adoptees were not included in formal government discourse concerning overseas Koreans until 1998, when President Kim Dae Jung issued an apology for the nation's international adoption participation. As "overseas Koreans," adoptees re-enter the Korean nation even though "they do not have any connection at all neither to Korea nor to things Korean, and nor to any overseas Korean community" (Hübinette 162).[3] The performative act thus provides adoptees new access to claiming a wholly Korean identity, whereby their adoptee status is marginal in relation to their existence as ethnic Koreans. Adoptees' online racial performance via the phrase "I am Korean American" intentionally disrupts dominant narratives of "Koreanness." Consequently, not only does featuring adoptees within IAmKoreanAmerican.com raise the profile of how international adoption shapes Asian America, it also serves as a forum to highlight the KAD nation.

In analysing how adult adoptees assert their "Koreanness" alongside their utterance of "I am Korean American," I gained insight into the ways in which marginalized communities insert their voices within diasporas. I root this inquiry in transnational feminism as it acknowledges how borders and boundaries framing identity categories are re-conceptualized and blurred (Friedman and Schultermandl 6). May Friedman and Silvia Schultermandl note how practices of international adoption provide a new lens to investigate identity categories in a globalized, transnational era (9). This research seeks to further existing conversations of diasporic populations in its "[reflection of] the transnational circulation of populations" (Grewal and Kaplan 16). By allowing ethnic Koreans to reterritorialize themselves online, I suggest that IAmKoreanAmerican.com melds Arjun Appadurai's two concepts: (1) mediascapes, "image-centered, narrative-based accounts" that "provide large and complex repertoires of images, narratives, and [world]" to inhabitants across the globe; and (2) technoscapes, which reflect how "technology, both high and low, both mechanical and information, now moves at high speeds across various kinds of previously impervious boundaries" (34–5). While Appadurai contends that mediascapes are tied directly to the visual, I find that to fully understand the significance of how IAmKoreanAmerican.com bridges territorialization disjunctures, we must consider how twenty-first-century technological

---

3 As part of this re-incorporation, adoptees gain access to the F-4 visa, which "allows adoptees, as overseas Koreans, to stay in South Korea for up to two years with rights to work, make financial investments, buy real estate, and obtain medical insurance and pensions" (Eleana J. Kim, "Wedding Citizenship and Culture" 59). The visa also exempts male adoptees from compulsory military service as well as disallows voting in elections (Hübinette 166). In April 2010, the South Korean government revised its Nationality Law to allow adoptees' access to dual citizenship, effective 1 January 2011.

advances (i.e., social media/social networking sites) influence the evolution of mediascapes. The site provides a sense of immediacy for members of the Korean diaspora's virtual world. By virtue of housing profiles of Korean Americans, the site offers the illusion that the diasporic community across the world is always connected. An ex-pat living in Europe or South America can submit his/her profile as easily as the *gyopo* (overseas Korean) teaching English in South Korea. No longer are Korean Americans and overseas Koreans, more broadly, isolated from one another. IAmKoreanAmerican.com serves as a forum to coalesce a collective identity that celebrates the heterogeneity within the diaspora.

Beyond the Declaration of "I Am Korean American"

Through the creation of a democratizing space (by allowing any individual to submit his or her profile for recognition), IAmKoreanAmerican.com deconstructs the term "Korean." For each participant, a central "Koreanness" is absent. Instead, the site produces counter-narratives, which, according to Homi Bhabha, "evoke and disturb those ideological maneuvers through which 'imagined communities' are given essentialist identities" (in Alarcón, Kaplan, and Moallem 7). The site celebrates the heterogeneity of the diaspora in its virtual ethnic neighbourhood. Participant profiles exist as mediated presentations of self, yielding new insights into how "everyday" adoptees trouble dominant notions of what it means to be Korean American. Due to the autobiographical nature of the "Own Words" section, I consider these snippets as a new form of autobiography, generated by the advent of Web 2.0 technology. The online profile, whether 300 characters or more as encouraged on the IAmKoreanAmerican.com submission page or constrained to 140 characters or fewer on Twitter, reflects how social media changes the way in which individuals share personal details of their lives with one another.

In reading the "Own Words" section as autobiography, I employ a feminist Asian American literary analysis. The profiles recall how the feminist autobiographical project, according to Leigh Gilmore, "is concerned with interruptions and eruptions, with resistance and contradiction as strategies of self-representation" (42). Sidonie Smith reinforces Gilmore's contention that autobiography is concerned with unearthing strategies of self-representation, providing a space for "restaging subjectivity" (159). Each narrative is a primary source document of an adoptee's own "telling." A feminist reading of these autobiographies provides an opportunity to better examine how adoptees reconcile the tensions in forming their identities as Asian Americans and contribute to the deconstruction of the "Asian American" identity (Cheung; Geok-Lim and Ling; Elaine Kim; Wong).

At the core of this inquiry is a commitment to explore adult adoptees' identity and the ways in which they construct understandings of self and belonging as citizens and diasporic subjects. While other scholars utilize ethnic and racial identity measures to capture adoptees' identity construction, it is my intent through a literary analysis to unearth how identity is fluid (Phinney). I argue that computations cannot quantify the fluidity of identity as respondents in these studies are asked to generalize their experiences (Hatchard; Wesolowski; Wickes and Slate). In limiting ourselves to measurable quantitative research, we risk pathologizing the adoptee experience and reducing it to a singular, linear narrative. As a result, I am particularly interested in exposing how adoptees' racial performance as Asians or Koreans is negotiated throughout their lives (Hübinette; McKee; Park Nelson). For the purposes of this chapter, I examine two macro-themes: (1) the absence of racial diversity in childhood and adolescence; and (2) reconciliation of racial, ethnic, and cultural identities.

The Absence of a Mirror, Limited Reminders of Difference

Even as adoptees actively assert a Korean ethnic identity via their performative utterance of "I am Korean American," their brief autobiographical details reveal the ways in which a positive understanding of what it means to be "Korean" was absent or at least muted during childhood and adolescence. Due to their limited interactions with other persons of Asian descent and the absence of role models, adoptees discuss how they never realized other adoptees also felt isolated in predominantly white suburbs or rural areas across the United States. Succinctly capturing the disjuncture between their racial physiognomy and cultural upbringing, Melissa F. writes, "Most of my life I have grown up being one of the only Asian kid that I knew, and many times even forgot that I was Korean till I actually looked in the mirror. I still have friends that don't believe I'm Asian since sometimes I pick up a southern drawl. (I pick up other accents very easily)." Her statement concerning forgetting that she was Korean illustrates how adoptees are assimilated into the white family and gain access to a culturally white identity. A "southern drawl" is implicitly linked to whiteness and incompatible with persons of Asian descent. For Melissa F., "being Asian" is linked to verbal performance and mannerisms.

At the same time, other adoptees recount how their racial identity was explicitly shaped by the lack of diversity in their environs. Megan W. writes, "I am a Korean American, although I don't feel Korean ... I was raised by a Caucasian family in a very Caucasian part of the country, the Midwest." Her not feeling Korean reflects how the adoptive family may erase racial difference, which reveals the assimilationist model deployed by many adoptive families in the late twentieth century to integrate the transracial, international adoptee into the household. By not acknowledging ethnic or racial difference, adoptees find themselves embracing the identities readily presented to them – cultural whiteness in this case, as the majority of adopters were white Americans. Echoing Megan W.'s sentiments, Samantha J. notes,

> I was born in South Korea and adopted by a nice American family when I was eight months old. I had a typical "American" upbringing as my family did not know much of Korean ways or traditions. I was raised with two brothers and a sister, one of whom, my younger brother, was also adopted from Korea. Growing up I often felt like an outsider. Culturally, growing up as a Korean in rural America was very difficult because I knew nothing of my history and ancestry. For the most part I adopted the ways of my American family until I got older and then began to explore my Korean heritage. Even though I always felt different, I never really felt "Asian" until I was in my 20's.

This narrative of identity negotiation echoes sentiments by adoptees in memoirs and documentaries and reflects adoptees' existence as "outsiders within" (Hill Collins; Trenka; Robinson; *First Person Plural*). Adoptees remain at the margins of American society because of race and the forever foreigner stereotype, while they are simultaneously insiders because of cultural whiteness (Twine). In addition, adoptees accessed the benefits of white privilege vis-à-vis their status as derivative citizens, which accounts for how their raced Asian bodies, historically, met the requirements for naturalization because of their adoptive parents' whiteness and American citizenship (Gardner). Nevertheless, Asian physiognomy prevents adoptees from gaining full white subjectivity, rendering them outsiders with an insider perspective of white cultural identity. Yet within the Asian American community, adoptees are simultaneously cast as outsiders because of their culturally white upbringing and as insiders based on physiognomy. Similarly, adoptees are rendered "outsiders within" upon return to Korea as a result of their lack of Korean cultural capital. Throughout their lived experiences, adoptees remain "outsiders within" precisely because they never gain full "insider" status in their multiple positionings.

Their status as outsiders within is significant as adoptees differentiate themselves from other Americans. This reflects how adoptees integrated a Eurocentric world view into their daily lives. To be "American" becomes equated with whiteness. As long as one's ethnic identity is tied to a racial construct outside of "white," the individual is not necessarily bestowed with the same privileges as other (white) Americans. For example, Melissa O. discusses the common question of "where are you from?," noting her response is, "I am basically Nebraskan." Her sentiment is common among adoptees featured on the site as they focus on the seemingly (white) Americanness of their lives. However, hesitancy is apparent by her use of the term "basically" to describe her statehood. The tensions produced by adoption are evident when adoptees discuss their intersectional identities. Almost but not quite white Americans, adoptees acknowledge their tenuous belonging. Recognizing the disjunctures between racial and cultural identities, Megan W. reflects, "Since I moved to California, I have been trying to embrace and learn about my Korean heritage." For her and Melissa O., a tangible *feeling of* being Korean is non-existent. Even as they may lack awareness of how stereotypes position persons of Asian descent as forever foreigners, adoptees at least have a basic understanding of how racial difference functions in their day-to-day interactions with individuals who operate with a white/black prism of race relations. Adoptees recognize that their positionality within their local communities (e.g., to be Nebraskan or from the Midwest) is implicitly linked to their adoptive families and adoption to the United States. These profiles reveal how the Internet continues to offer new methods for identity exploration. Interrogating one's ethnicity and meeting other adoptees are enhanced through online media such as IAmKoreanAmerican.com. Through revealing the complications of what it means to be American, adopted, and Korean in their autobiographies, these individuals may validate the feelings of other adoptees who access the site. In other words, profile viewers may reach an "aha!" moment, recognizing pieces of themselves or their experiences in the lives of these individuals.

The isolation experienced by adoptees in childhood and adolescence is reflective of how, for many, an active negotiation of what it means to be ethnically Korean was not undertaken until adulthood. In this regard, Jamey B. writes, "I hated being Korean for the first 22 years of my life. I am now very proud to be Korean." For Jamey B. the concept of "Koreanness" and its innate meaning affected how he saw himself. The visible racial difference for adoptees in predominantly white geographic locations was exacerbated by a lack of positive representations of Asian Americans in the media. For these adult adoptees, access to the Internet and sites like IAmKoreanAmerican.com would have been critical in their identity development. Social media has fundamentally changed the ways in which adoptees from more recent generations have connected and sustained relationships with other adoptees. This particular site serves as an alternative to geographic isolation from other adoptees and Koreans. In a study of 23 Korean adoptees living in Minneapolis–St. Paul, Minnesota, Dani Meier discovered that only when adoptees left for university did they first experience a heightened level of diversity because during childhood they grew up in predominantly white communities. He found that in adulthood ethnicity was more salient because adoptees were visibly different from the dominant population, even though the Twin Cities has a thriving adult Korean American adoptee community.[4] A lack of racial diversity also has an impact on adoptees' cultural socialization, "the ways in which parents negotiate the racial, ethnic, and cultural experiences within the family and seek to promote or hinder racial and ethnic identity development in the child" (Song and Lee 22).

Racial salience shifts because of adoptees' varying levels of ethnic exploration – "the personal examination of one's ethnic ancestry and its relevance for one's life" (Shiao and Tuan 1024). Jiannbin L. Shiao and Mia H. Tuan note, "When adoptees leave home in early adulthood, they also leave an important anchor of their status as honorary whites, while their status as nonwhites remains unchanged and may increase in salience" (1030). Reconciling the complexities of identities, adoptees as outsiders within also re-examine who they are in response to return to Korea or participation within the wider Asian American community. Young Sook P. comments, "Yet, there are times I feel so cut off from my culture. Being in Korea made me

---

4 This includes a local adult Korean adoptee organization, AK Connection.

realize the disconnect more. I'm sure that other Korean adoptees can relate – being 'Korean' but not 'feeling Korean.' It's a struggle with identity, to try to capture the place where you fit." Young Sook's reflections are not uncommon. Rather, in memoirs and anthologies Korean adoptees continuously ruminate over the cultural competency required to "be" fully Korean (Bishoff and Rankin; Cox). These individuals also consider the dissonance between operating within Korea and the ways in which they are viewed as stereotypically Asian in the United States.

Embracing the Contradictions: The Intersection of Identities

Even as a disconnect is evident in how adoptees label themselves between childhood and adulthood, the reconciliation of their multiple racial, ethnic, and cultural identities does not simply end when they begin viewing themselves as Korean American. Rather, the complexities of identity are revealed. Examining her shift in identity, Karen H. finds, "Now that I've begun the process of exploring my Korean heritage a lot of identity questions have come up. What does it mean and look like to be Korean American? How do I embrace my Korean side without diminishing the American side ... and the values my parents taught me?" While other scholars have discussed how such dissonance is a sign of racial melancholy, I suggest that the dualities adoptees inhabit reflect the "in-between" spaces of trans-nationalism (Eng; S. Park; Bhabha). Their positioning within the "in between" is clearly evident when considering their experiences returning to South Korea. Reflecting on her first return visit in 2009, Sue R. notes,

> I often struggle to answer the question, "Where are you from?" Despite growing up in America, I cannot help but reflect on my experiences in Korea and consider the argument for nature vs. nurture. Am I product of my environment? My genetic make-up? Can I consider both elements? On the outside I appear Korean, but my soul speaks American. I've realized one's categorization of themselves can fluctuate and people don't have to isolate their home as solely "Korea" or "America." Experiences take us across multiple borders and cultures, ultimately changing our perception of the world and ourselves. By answering, "Korea" and "America" we allow ourselves to accept and embrace these wonderful cultural characteristics particular to each country.

Her intimate reflection on how transnational adoption influences her sense of self illustrates adoptees' outsiders within status. Recognizing her lack of Korean cultural competency, she examines how the terms "Korean" and "American" are not static. To be Korean or American is not an either/or dichotomy. Embracing the contradictions, Sue R.'s statement illustrates how adoptees may struggle with their sense of national and cultural belonging. Adoptees continuously engage a fluid sense of Asian American identity. Troubling what it means to be authentically Korean or American, adoptees featured on IAmKorean-American.com carve a new space for themselves within conceptualizations of "Korean American." The site offers a forum for adoptees to actively claim their place within the Korean diaspora with the click of the submit button. Adoptees' performative utterance secures inclusivity in a once exclusionary discourse that overlooked them as Korean subjects. Thus adoptees negotiation of identity cannot only be seen as a personal undertaking. Rather, through participating in the site, adoptees are also involved in a greater shift requiring a rethinking of static notions of identity.

Yet the dichotomy of "Korean" or "American" is evident in the lives of adoptees. This binary is revealed in their discussion of consumption of Korean cuisine and stereotypes. Their appeal for authenticity relies on adoptees drawing from historical tropes concerning Asians as bad drivers, superior at math compared to other racial groups, and loving spicy food. For example, Melissa O. writes, "I guess I am your typical Asian in some ways: bad driver, fan of all *Fast and the Furious* movies, techno/trance is some of the best music out there, and I guess you could say I'm like a human calculator. But then again I have two tattoos and hoping for many more and no I'm not going to be a doctor but rather a chef." Likewise, Michelle K. recounts, "I have no idea what being Korean American is all about or what it means or what the implications are. But, I love spicy food and I've come to learn that there is no problem too large that cannot be solved with tons of drinks, crazy friends and a karaoke room. This was taught to me by Korean Americans." The East-versus-West juxtaposition of food, entertainment, and stereotypes shapes how adoptees' interpret their likes and dislikes. The belief that American and Korean cultures remain static and cannot be fused together is also evident in Valissa P.'s reflection on her adoption experience:

I was adopted at 3 months old from Seoul, South Korea. I grew up in a mostly Caucasian, suburban environment, but had many friends and cousins who were adopted not only from Korea, but from various different countries. It wasn't until my early twenties that I expressed interest in learning more about Korean culture. I befriended several Koreans, participated in tae kwon do, learned how to make japchae, and even attempted to learn a little bit of the language. Although I have transitioned well into American culture, there is still a large part of me that will always remain Korean. This was difficult for me to comprehend growing up, but now I realize it is an asset. Not many people have the opportunity to "belong" to two countries. In the near future, I desire to travel to Seoul and experience my homeland firsthand.

By constructing identity in fixed terms, whereby Korean culture is linked to the consumption of specific cuisine, language, and sport, Valissa P. clearly delineates between East and West. She also conflates racial and national identities, with the assertion that "part of [her] will always remain Korean." This emphasis on distinct cultures and experiences is implicitly linked to how stereotypes of Asian Americans penetrate and circulate within the United States. With a limited number of positive representations of Asian Americans throughout the twentieth century and perhaps few intersections with persons of Asian descent, adoptees may construct their perceptions of the Asian Other vis-à-vis Orientalist scripts in media.

Reconciling their status as outsiders within, their profiles reveal how identity is not easily formed and is a process in constant negotiation. Adoptees participate in a wider discussion of intersectionality and the ways in which issues of power and privilege operate in their lives (Anchisi; Pearson). Epitomizing how one's intersectional identity may constrain how adoptees define themselves, Andy M. asserts,

> I am Korean American. I am an adoptee. I am a transgender woman. It would seem that I am also a conundrum. Some would argue that I am terribly challenged. But my identity is a prime reason for why I became an activist. I became an activist so that I could liberate myself from the stereotypes surrounding my identity.

While she delineates between her multiple "check-boxes" of identity, Andy M. recognizes the ways in which these "boxes" may limit her ability to enact agency. The conundrum of an intersectional identity requires individuals' self-reflexivity and ability to negotiate the complexities of their outsiders within status. With the recognition that stereotypes may inhibit how she is perceived and received by the world, Andy M.'s declaration of "I am Korean American" asserts how the processes of racialization experienced by Asian Americans more broadly are negotiated at the individual level. She consciously negotiates her positionality, careful to invoke her intersectional identity to ensure her experiences are not marginalized.

In the assertion of a Korean American identity, adoptees also construct their identities as a political statement. Using the "Our Words" section as a forum to critique international adoption, Sun Yung S. contends, "Being Korean American means many things to me – as a transnational adoptee it means legal abandonment, a police station, a social worker, a foster home, an orphan hojuk, two passports, two birth certificates, a naturalization certificate, several names." Tracing how she became part of the wider Korean diaspora, Sun Yung S. conjures the history of transnational adoption for the reader. Highlighting the ways in which adoptees unwillingly engaged in migration and separation to become Korean Americans, her reflections should be viewed within broader debates that challenge the portrayal of adoption as an act of humanitarianism. Echoing this criticism, Kristin P. writes,

> Like 10% of Korean Americans, I am an involuntary immigrant. I was brought to the USA to be the daughter of a white American couple. I had no say in becoming American, whether or not I now embrace, reject, or simply accept that this is my identity. I do accept it, both aspects. Some adoptees feel like they're not Korean but through my friendships with other 1.5/2nd generation Korean Americans, I see that I'm on the continuum of a huge range of what it means to be Korean American.

The insertion of their migration histories to the United States not only allows Sun Yung S. and Kristin P. to claim a Korean American identity, but also provides the opportunity to challenge dominant narratives of adoption. In doing so, the pair reconstruct the origins of adoptees within the diaspora and highlight the ways in which adoptees were once previously cast from the nation. The inclusion of their unmediated constructions of the adoptee community emphasizes how IAmKoreanAmerican.com democratized whose voices are included in histories of adoption or in the Korean diaspora since the site did not censor autobiographical content. The Web 2.0 profiles offer unfettered access to what it means to be a transnational subject.

Conclusion

I used to have lots of existential angst over how "Korean" or not I am ... for now, I'm comfortable telling people I'm Korean American ... but that labels are just that – labels, and really, I am more complex than a box on a piece of paper.

Lena Soo Hee W., 6 August 2010

In analysing adoptees' self-presentation, this study contributes to a growing body of literature that examines and interrogates the ways in which adoptees represent themselves in print and visual media. I demonstrate how adopted Koreans reinvent the standard definition of the phrase "I am Korean American." Their performative utterance illustrates how adoptees are, as Lena Soo Hee W. writes, "more complex than a box on a piece of paper." While adoptee profiles may parallel one another through repeated keywords and themes, their lived recollections and engagement with the Korean American community vary. This heterogeneity is readily apparent when considering Nicholas O., who writes, "Just discovered the site and thought it would be interesting to post a profile. I've never really connected with my heritage since I was adopted at a few months old from Seoul. My parents are white, hence the last name. I would like to travel back to South Korea again now that I am an 'adult.'" Yet through submitting his profile, Nicholas O. asserts his performative ability to claim a Korean American identity. In many ways, he embodies the concept of the outsider within, given the disjunctures between his cultural competency as Korean and his ethnicity. Exposing the multitudes of adoptee voices found within the ethnic website, this chapter documents the way in which a singular diaspora can no longer exist. Rather, a transnational feminist approach unearths the heterogeneity found within a singular diasporic constituency as well as in the wider diaspora at large.

The Internet serves as an extension of the Korean adoptee imagined community, allowing adoptees the ability to construct their public identities – how they perceive themselves and how they want others to perceive them. Critically interrogating the use of the Internet, adoptees forge and intervene in dominant constructions of the "Korean American" identity in their deployment of an online "Korean" persona. Publishing information about themselves for immediate public consumption, the site's "Own Words" section provides snippets of the lives of adoptees and their self-representation as Korean Americans. These autobiographies reflect the many ways social media changes how individuals effectively communicate asynchronously across the Internet. The rapidity in which Web 2.0 media has allowed for connections to be created and sustained cannot be overlooked. Once deterritorialized and sometimes isolated communities are finding themselves in growing imagined communities. At the same time these new technologies are also allowing for the growth of a new form of non-biogenetic kinship, including adoptees' contingent essentialist ties.

WORKS CITED

Alarcón, Norma, Caren Kaplan, and Minoo Moallem. "Introduction: Between Woman and Nation." *Between Woman and Nation: Nationalisms, Transnational Feminisms, and the State.* Ed. Caren Kaplan, Norma Alarcón, and Minoo Moallem. Durham, NC: Duke University Press, 1999. 1–16. Print.

Anchisi, Lidia. "One, No One, and a Hundred Thousand: On Being a Korean Woman Adopted by European Parents." *The Intersectional Approach: Transforming the Academy through Race, Class, and Gender.* Ed. Michele Tracy Berger and Kathleen Guidroz. Chapel Hill: University of North Carolina, 2009. 290–9. Print.

Appadurai, Arjun. *Modernity at Large: Cultural Dimensions of Globalization.* Minneapolis: University of Minnesota, 1996. Print.

Baym, Nancy K. "The Emergence of On-Line Community." *CyberSociety: Computer-Mediated Communication and Community.* Ed. Steven G. Jones. Thousand Oaks, CA: Sage, 1996. 35–68. Print.

Bhabha, Homi. *The Location of Culture.* 1994. London: Routledge, 2004. Print.

Bishoff, Tonya, and Jo Rankin, eds. *Seeds from a Silent Tree: An Anthology by Korean Adoptees.* San Diego: Pandal, 1997. Print.

Butler, Judith. *Bodies That Matter: On the Discursive Limits of "Sex."* New York: Routledge, 1993. Print.

– *Gender Trouble: Feminism and the Subversion of Identity.* New York: Routledge, 1990. Print.

Byom, Jamey. "I Am Korean American." *I Am Korean American.* Barrel, 10 Nov. 2009. Web. 9 Sept. 2013. <http://iamkoreanamerican.com/2009/11/10/jamey-byom/>.

Chen, Tina. *Double Agency: Acts of Impersonation in Asian American Literature and Culture.* Stanford: Stanford University Press, 2005. Print.

Cheung, King-Kok, ed. *An Interethnic Companion to Asian American Literature.* Cambridge: Cambridge University Press, 1997. Print.

Cox, Susan Soon-Keum. *Voices from Another Place: A Collection of Works from a Generation Born in Korea and Adopted to Other Countries.* St Paul, MN: Yeong & Yeong Book, 1999. Print.

Danet, Brenda. "Text as Mask: Gender, Play, and Performance." *CyberSociety: Computer-Mediated Communication and Community.* Ed. Steven G. Jones. Thousand Oaks, CA: Sage, 1996. 129–58. Print.

Eng, David L. *The Feeling of Kinship: Queer Liberalism and the Racialization of Intimacy.* Durham, NC: Duke University Press, 2010. Print.

Feroe, Melissa. "I Am Korean American." *I Am Korean American.* Barrel, 8 Apr. 2010. Web. 7 Sept. 2013. <http://iamkoreanamerican.com/2010/04/08/melissa-feroe/>.

First Person Plural. Dir. Deann Borshay Liem. Center for Asian American Media, 2000.

Friedman, May, and Silvia Schultermandl. "Introduction." *Growing Up Transnational: Identity and Kinship in a Global Era.* Ed. May Friedman and Silvia Schultermandl. Toronto: University of Toronto Press, 2011. 3–18. Print.

Gardner, Martha. *The Qualities of a Citizen: Women, Immigration, and Citizenship, 1870–1965.* Princeton, NJ: Princeton University Press, 2009. Print.

Geok-Lim, Shirley, and Amy Ling. *Reading the Literatures of Asian America.* Philadelphia: Temple University Press, 1992. Print.

Gilmore, Leigh. *Autobiographics: A Feminist Theory of Women's Self-representation.* Ithaca, NY: Cornell University Press, 1994. Print.

Grewal, Inderpal, and Caren Kaplan. *Scattered Hegemonies: Postmodernity and Transnational Feminist Practices.* Minneapolis: University of Minnesota Press, 1994. Print.

Harms, Karen. "I Am Korean American." *I Am Korean American.* Barrel, 27 July 2010. Web. 9 Sept. 2013. <http://iamkoreanamerican.com/2010/07/27/karen-harms/>.

Hatchard, Christine Jung. "Racial Experiences of Korean Adoptees: Do Adoptive Parents Make a Difference?" Diss., Chestnut College, 2007.

Hill Collins, Patricia. *Black Feminist Thought: Knowledge, Consciousness, and the Politics of Empowerment.* New York: Routledge, 2000. Print.

hooks, bell. *Feminist Theory from Margin to Center.* 2nd ed. London: Pluto, 2000. Print.

Hübinette, Tobias. *The Korean Adoption Issue between Modernity and Coloniality.* Sarrbrücken, Germany: Lambert Academic, 2009. Print.

Jo, Sunny. "Making of KAD Nation." *Outsiders Within: Writing on Transracial Adoption.* Ed. Jane Jeong Trenka, Julia Chinyere Oparah, and Sun Yung Shin. Cambridge: South End Press, 2006. 285–90. Print.

Johnson, Samantha. "I Am Korean American." *I Am Korean American.*
Barrel, 13 Apr. 2010. Web. 9 Sept. 2013. <http://iamkoreanamerican.
com/2010/04/13/samantha-johnson/> (site no longer available).

Kendall, Lori. "Recontextualizing 'Cyberspace': Methodological
Considerations for On-Line Research." *Doing Internet Research: Critical Issues
and Methods for Examining the Net.* Ed. Steve Jones. Thousand Oaks, CA:
Sage Publications, 1999. 57–74. Print.

Kim, Elaine H. "Defining Asian American Realities Through Literature."
Cultural Critique 6 (1987): 87–111. Print.

Kim, Eleana J. "Remembering Loss: The Cultural Politics of Overseas
Adoption from South Korea." Diss., New York University, 2007.

Kim, Eleana J. "Wedding Citizenship and Culture: Korean Adoptees and the
Global Family of Korea." *Social Text* 21.1 (2003): 57–81. Print.

Koehn, Michelle. "I Am Korean American." *I Am Korean American.*
Barrel, 20 Oct. 2010. Web. 9 Sept. 2013. <http://iamkoreanamerican.
com/2010/10/20/michelle-koehn/>.

Marra, Andy. "I Am Korean American." *I Am Korean American.* Barrel, 12 Mar.
2010. Web. 9 Sept. 2013. <http://iamkoreanamerican.com/2010/03/12/
andy-marra/>.

McKee, Kimberly. "Real versus Fictive Kinship: Legitimating the
Adoptive Family." *Critical Kinship Studies: Kinship (Trans)Formed.*
Ed. Charlotte Kroløkke Lene Myong, Stine W. Adrian, and Tine
Tjørnhøj-Thomsen. London: Rowman and Littlefield International,
forthcoming. Print.

Meier, Dani I. "Cultural Identity and Place in Adult Korean-American
Intercountry Adoptees." *Adoption Quarterly* 3.1 (1999): 15–48. Print.

Mitra, Ananda. "Using Blogs to Create Cybernetic Space: Examples from
People of Indian Origin." *Convergence: The International Journal of Research
into New Media Technologies* 14.4 (2008): 457–72. Print.

Nakamura, Lisa. *Cybertypes: Race, Ethnicity, and Identity on the Internet.* New
York: Routledge, 2002. Print.

Offner, Melissa. "I Am Korean American." *I Am Korean American.*
Barrel, 16 Apr. 2010. Web. 9 Sept. 2013. <http://iamkoreanamerican.
com/2010/04/16/melissa-offner/>.

Osgood, Nicholas Lee. "I Am Korean American." *I Am Korean American.*
Barrel, 5 Jan. 2011. Web. 9 Sept. 2013. <http://iamkoreanamerican.
com/2011/01/05/nicholas-lee-osgood>.

Pak, Kristin. "I Am Korean American." *I Am Korean American.* Barrel, 26 Jan.
2010. Web. 9 Sept. 2013. <http://iamkoreanamerican.com/2010/01/26/
kristin-pak>.

Park, Shelley M. *Mothering Queerly, Queering Motherhood: Resisting Monomaternalism in Adoptive, Lesbian, Blended, and Polygamous Families.* Albany: State University of New York Press, 2013. Print.

Park, Young Sook (Kate). "I Am Korean American." *I Am Korean American.* Barrel, 10 May 2010. Web. 9 Sept. 2013. <http://iamkoreanamerican. com/2010/05/10/young-sook-kate-park/>.

Park Nelson, Kim. "Korean Looks, American Eyes: Korean American Adoptees, Race, Culture and Nation." Diss., University of Minnesota, 2009.

Pearson, Holly. "Complicating Intersectionality through the Identities of a Hard of Hearing Korean Adoptee: An Autoethnography." *Equity & Excellence in Education* 43.3 (2010): 341–56. Print.

Perry, Valissa. "I Am Korean American." *I Am Korean American.* Barrel, 7 Dec. 2010. Web. 9 Sept. 2013. <http://iamkoreanamerican.com/2010/12/07/valissa-perry/>.

Phinney, Jean S. "Ethnic Identity in Adolescents and Adults: Review of Research." *Psychological Bulletin* 108.3 (1990): 499–514. Print.

Rissberger, Sue. "I Am Korean American." *I Am Korean American.* Barrel, 8 Sept. 2010. Web. 9 Sept. 2013. <http://iamkoreanamerican. com/2010/09/08/sue-rissberger/> (site no longer available).

Robinson, Katy. *A Single Square Picture: A Korean Adoptee's Search for Her Roots.* New York: Berkley, 2002. Print.

Shiao, Jiannbin Lee, and Mia H. Tuan. "Korean Adoptees and the Social Context of Ethnic Exploration." *American Journal of Sociology* 113.4 (2008): 1023–66. Print.

Shin, Sun Yung. "I Am Korean American." *I Am Korean American.* Barrel, 22 Mar. 2010. Web. 9 Sept. 2013. <http://iamkoreanamerican. com/2010/03/22/sun-yung-shin/>.

Smith, Sidonie. *Subjectivity, Identity, and the Body: Women's Autobiographical Practices in the Twentieth Century.* Bloomington: Indiana University Press, 1993. Print.

Song, Sueyoung, and Richard Lee. "The Past and Present Cultural Experiences of Adopted Korean American Adults." *Adoption Quarterly* 12.1 (2009): 19–36. Print.

Trenka, Jane Jeong. *The Language of Blood: A Memoir.* St Paul: Graywolf, 2003. Print.

Tuan, Mia. *Forever Foreigners or Honorary Whites? The Asian Ethnic Experience Today.* New Brunswick, NJ: Rutgers University Press, 1998. Print.

Turkle, Sherry. *Life on the Screen: Identity in the Age of the Internet.* New York: Simon & Schuster, 1995. Print.

Twine, France Winddance. "Brown Skinned White Girls: Class, Culture and the Construction of White Identity in Suburban Communities." *Gender, Place and Culture: A Journal of Feminist Geography* 3.2 (1996): 205–24. Print.

Wesolowski, Paul Kim. "Ethnic Identity Development of Korean, International, Transracial Adoptees." Diss., Wright Institute, 1996.

Wickes, Kevin L., and John R. Slate. "Transracial Adoption of Koreans: A Preliminary Study of Adjustment." *International Journal for the Advancement of Counseling* 19 (1996): 187–95. Print.

Wilkinson, Hei Sook Park, and Nancy Fox, eds. *After the Morning Calm: Reflections of Korean Adoptees.* Bloomfield Hills, MI: Sunrise Ventures, 2002. Print.

Wilson, Samuel M., and Leighton C. Peterson. "The Anthropology of Online Communities." *Annual Review of Anthropology* 31.1 (2002): 449–67. Print.

Wolters, Megan. "I Am Korean American." *I Am Korean American.* Barrel, 22 July 2010. Web. 9 Sept. 2013. <http://iamkoreanamerican. com/2010/07/22/megan-wolters/>.

Wong, Sau-Ling C. "Immigrant Autobiography: Some Questions of Definition and Approach." *American Autobiography: Retrospect and Prospect.* Ed. Paul John Eakin. Madison: University of Wisconsin, 1991. 142–70. Print.

Wood, Lena Soo Hee. "I Am Korean American." *I Am Korean American.* Barrel, 6 Aug. 2010. Web. 9 Sept. 2013. <http://iamkoreanamerican. com/2010/08/06/lena-soo-hee-wood/>.

Yuh, Ji-Yeon. *Beyond the Shadow of Camptown: Korean Military Brides in America.* New York: New York University Press, 2004. Print.

Disembodied Connections

8 Shifting Terrain: Exploring the History of Communication through the Communication of My History

MAY FRIEDMAN

Introduction

My family history is characterized by migration and upheaval. My parents are living in their third country, speaking their third language. All of my children's grandparents were displaced and relocated. The majority of my enormous family lives in a different place from me. As a result, my family history has been quilted together through communication technologies, some very rudimentary and others quite contemporary, and through the "complex and nuanced manifestations of transnational circulations of peoples, goods, and information in the present moment" (Alarcón, Kaplan, and Moallem 4). Understanding my history is as much a story of communication as a story of migration, transnationality, and kinship.

It is tempting to consider the history of communication and the family as a linear path. In the past, communication was episodic and expensive; airmail letters would cross en route and thus result in a dialogue that was out of sync. Long-distance telephone calls were prohibitively expensive and thus constrained communication to only the most essential messages. By contrast, both the speed and the virtual embodiment of "modern" communication avoids the great gap between communiqués, bridges the distance between family "there" and "here," and allows for a means of connection that provides us multisensory access to one another. Yet there are losses here, too. Despite our capacity to connect, our over-connection may result in a busyness that limits the time for reaching out. Language limitations, direct consequences of migration, are not necessarily mitigated by new technologies. And the real-time dialogue which new media allows may nonetheless be as flawed and inauthentic an archive as the missives which preceded it. By exploring particular moments in the history of communication and the history of my family, I aim to consider the implications of communication, transnationality, and kinship. Who do we become as we communicate? Who is left behind in the words and images that remain?

Studies of communication have often focused on mass communication and the ways that shifts in communication culture have shifted movements and policies (Youngs; Hegde; Shohat and Stam). In this chapter, I aim to expose micro communications, the means by which a family is pushed together and pulled apart. By focusing on the themes of identity, dialogue, language, and archive, I will examine the history of my own family as one example of the effects of shifting communication technologies on individual and familial identities as well as on national and transnational subjectivities.

Identity

My dear Shola. I'm writing this letter to try to write with an English. I hope that to help us to learn the language. If you also write to me with an English. We learn the English language more quick.

Before two days I send a post card with an English to you. I am expecting from you to understand what I meaning my wonderful wife. I know! That is the hard language for our loves letters. We must not alarm my angel. I love you always and all languages aren't enough for us. Kiss to Jhodit and see you –

I was around 12 when I found the letters. Snooping around in my mother's drawers, looking for tights or lipstick, I happened upon an overflowing shoebox filled with letters from my parents' time apart. Initially separated by my father's military duty, my parents were then parted by the process of immigration. My father spent almost a year in Canada, getting organized, before bringing over my mother and sister. These letters are miraculous representations of a bygone era, letters that document both the horror of my father's wartime experiences and the hopefulness and loneliness of immigration to a new land. Or at least, that's what I think they represent. Because they're almost entirely not in English, I can't actually read the letters. I can pore over the dates, imagine what they must say, look at the bizarre images of "Canada" (circa 1969) on the backs of the postcards, but I can't actually access my parents' younger selves. The very few letters in English show my father's commitment to move to Canada and his motivation to use English instead of their mutual native tongues even when it was not required, as a means of mastering their new language. My mother's letters show her ambivalence about leaving, about English, even as she sternly corrects his grammar.

My lovere, my dear husband! I love you and wait for day I see you. And I hop it will be shortly. I understand almost all your letter. But you must to write "told him" no "told hem." On this Saturday I was at my parents in Kadima. My sister Dalia very angre you, because you do not come back. She want to see you very much. I did not answer her when she talking. I do not want speak about it because it very difficult for me to leave my family and go with you but I want see you and to life with you all my life days. So, how I can to stay her and you there?

We are mere decades beyond those letters, yet the passage of time, for my family and for communication and media, have rendered these missives quaint, archaic. I wonder how much of their essential selves can be found in these letters. When I remember my father, do I remember his hopeful 30-something self, carefully writing in childish English cursive? Do I remember paging him on his taxi radio in my own childhood? Or the ubiquity of his present-day cell phone, binding him (with some reluctance) to family as he keeps working and driving into his late seventies?

If our identities are performances, discursively constructed (Butler), our communications with our loved ones are some of the most visible manifestations of those performances, mediated by the instruments that we use. The three languages of my family reflect our three national birthplaces – Iraq, Israel, and Canada – and reflect both our travels and the virtual borders between us. Beyond language, our family represents a technological revolution, with each member, including my own children, outstripping the technical know-how of his or her elders. My identity is thus hybrid (Lemke) and shape-shifting, comprising the academic prose I carefully construct, the short texts I send to my sister and the multilingual phone calls with my mother, the in-person hollering dialogue with my father (who has been hard of hearing for my entire life), with endless other interactions with family and media interspersed. We are all performing our identities in relation to one another (Shohat; Mohanty; Shohat and Stam), but also in relation to the technologies on which we have come to depend.

While our experiences of these technologies can sometimes be invisible, they are not necessarily benign. There is great potential for our capacity to mediate family at a distance, but it is arguable that we are creating different selves as a result. The new subjectivities which are revealed as a result of technological mediation are not, in and of themselves, positive or negative, but they require a degree of self-reflection in order for us to understand the consequences of new communication, a commitment to considering "the story of the telling of the story" (Simon, Rosenberg, and Eppert 7). Theories of life writing ask us to consider the ways that a life narrated, through autobiography, memoir, or letters, can allow for the authorship of a "self-in-the-writing" (Kadar 12). As the explosion of new communication technologies allows us to interact, what are the implications for the selves we are making, especially in the realm of interfamilial and transnational communications, in the intimacy and mundanity of family chat, mediated through the distance and intricacies of global connections?

I do not mean to sound like a Luddite or a technophobe. I am a huge fan of new media and, indeed, have argued strenuously for the ways that access to new media has contributed to my own subject formation (and re-formation) (Friedman). Indeed, much of the collaboration that led to this volume of writing maximized the potential for Skype and other embodied forms of communication to link us across time zones and vast geographic distances. While I passionately embrace new media, however, I am nonetheless cognizant of the ways that it has shifted our identities, as well as the spaces between them.

In the story of my family, I see that our construction of ourselves, individually and collectively, has shifted as a result of our rapid path through letters, phone calls, emails, Skype, and Facebook: our identities are, to quote Stuart Hall, "a matter of 'becoming' as well as being" (225). I likewise see that these technologies are unevenly applied, partially because of the digital divide (Youngs) and partially because of the geographic, linguistic, and ability-based limitations that hinder familial communications. As Shohat argues, we exist "not as hermetically sealed entities but, rather, as part of a permeable interwoven relationality" (2). As families, perhaps particularly families interrupted by distance, we have always had an imperfect reckoning of one another. New media, however, presents a different set of imperfections than the non-technological communications of our past. In other words, my parents' love story isn't "the truth" of their lives: even the small subset of their writing which I can access in English makes it clear that they are choosing their words carefully, presenting themselves in particular ways. The digitized communications we now share with family members "back home" is likewise constrained. Of relevance, however, is that the performance has shifted, has resulted in a different set of truths revealed and occluded than the stories of ourselves as subjects "before." In order to further explore the effects of identity construction on subject formation, it is important to consider the implications of authenticity and the notion of a dialogic construction of truth.

Authenticity and Dialogue

Migration and communication have literally shifted my family's identities in calling attention to their very names. In my parents' letters they have not yet committed to the English spellings of their names and thus try many different versions of the spellings on for size. I am accessing not only my mother's younger self but also some strange version of "Shola" who does not resemble the mother, Shula, whom I have come to know; my father, "Uzy," is likewise distinct from my father, who on immigrating to Canada reclaimed his formal Arabic birth name, Aziz. That each of my father's brothers and their children have chosen different spellings for the family name was known to me, but that my father initially went by "Dlomy" before settling on "Dalume" was not. While these name changes may seem like trite examples of the implications of self-definition, communication, and transnationalism, they are reminiscent of the ways that "[g]lobalization represents a complex disjunctive order that is clearly not captured by the popular rhetoric of easy fusion and smooth cultural transitions; rather, it is marked by jagged contours which can no longer be captured in terms of simple binaries that characterize center-periphery models" (Hegde 1–2).

The process of migration may not have concretized my parents' identities, but it shifted their literal legal standing: upon arriving in Israel, each of my mother's sisters assumed a more "Zionized" name, abandoning the Arabic names they had borne to that point; on entering Canada, the Hebraicized names for my mother and sister were likewise shifted to the names they now irrevocably bear on all their documentation. Thus my sister (referred to in the letters as "Jhodit") went from the Hebrew name Yehudit to Judy, the name she uses today. My family's realities have been performed in relation to state apparatuses of citizenship and migration, honed through communications with one another and concretized through the filing of formal identity papers that may, nonetheless, erase aspects of their identities. These shifts of identity are unambiguous, but are they meaningful?

As a family displaced by colonialism, violence, war, and the arguably equally potent violence of immigration and racism, we learn from the story we tell. Simon, Rosenberg, and Eppert argue that we learn through our "lessons of the past," especially when those lessons include conditions of trauma. They argue that "all formations of memory carry implicit and/or explicit assumption about what is to be remembered, how, by whom, for whom, and with what potential effects" (2). Indeed, some argue that it is imperative that we begin to decode our own histories and their interrelationships with media and technology; Shohat and Stam insist that "transnationalizing media studies has become a political and pedagogical responsibility" (5).

I am seven years old and visiting with my mother. The wait to have a telephone installed in the small impoverished town where my aunts live, the town my mother left, is seven years long. As a result, when my father wants to call us from Canada, he phones the one neighbour lucky enough to have a phone, and she bellows out the window into the central courtyard that the apartment buildings share. We sit there to avoid the worst of the afternoon heat, telling stories and playing cards, when we hear her cry and my mother springs to her feet and runs up the two flights of stairs. While waiting for my mother, the neighbour chats with my father casually, but my mother, mindful of the prohibitive long-distance charges, quickly updates him: yes, she has seen his mother; no, his aunt isn't feeling better; yes, I am fine; no, my sister isn't enjoying herself.

Fast-forward 25 years: my six-year-old cousin sits in the same courtyard, studiously playing a game on an old iPhone. Another cousin phones me from 20 feet away to tell me to join him, that he doesn't want to move to get my attention. The ubiquity of communication culture is stunning: mere years after the last few landlines are finally installed, households routinely have five or six cell phones, several per member; even my poorest family members have a way to call me.

If our selves are constructed and mediated by distance and technology, then how can we access the authentic truth of our existence? This is a facile question, of course: there is no empirical subjectivity that we are masking. Instead, we are putting forth different selves at different moments and quilting together a patchwork subject based on our interactivity with tools and people. Thus, my parents' letters tell me a story that rests somewhere between his words and hers, between my father's optimistic version of events and my mother's reluctant immigration.

Likewise, the story of how we are doing on our trip is mediated through my mother's clipped narration but also in the interaction between my father and the loquacious neighbour. In the present era, this dialogic communication is increasingly disembodied. Stories are told and retold through email threads, Facebook chains, Twitter messages, in ways that allow us to create communicative artefacts that are marked by every interaction and retelling. As a result, the primary means of our communication has grown increasingly dialogic.

The letters sent to my grandparents often included snapshots, of my awkward Afro and giant teeth, of my sister beaming at graduation, of my father in front of Niagara Falls. These snapshots tell a story of our lives in Canada. The letters themselves, however, are also snapshots, limited snippets of our lives. Even if there were responses to these moments, they were so vastly temporally distant, occurring weeks after the letters were penned, that the dialogue they constructed was always off-kilter and awkward. By contrast, I can now watch my baby cousin take his first steps mere moments after this event, despite the 7,000 kilometres that separate us. I can store long email chains of conversation with family members, peppered by memes and YouTube clips that provide a history of sorts of our family's bifurcated and transnational existence. When I cannot be there, I can nonetheless experience some version of "there" in ways that my mother's hasty updates to my father, with the neighbour lurking nearby, may have lacked.

Perhaps the single biggest difference between the communication technologies of before and those we live with now is speed. While family members are increasingly dispersed across the globe, the capacity for instant and sustained communication has exploded. Thus my mother's weekly phone call to her sisters need no longer be undertaken at 7:00 a.m., when the rates were low, and can assume the kind of rambling dialogue that includes three or four follow-up calls to include forgotten information, a kind of interaction that has become indiscernible from our local communication.

We can interact with one another in real time, through our voices, our words, and our bodies. Indeed, a key difference from the past is the degree to which modern communication has become embodied. I imagine my father writing "I very yearn for you" in 1969, unable to imagine the brave new world where, denied my mother's embrace, he could at least hear her voice and see her smile. My children relate to their paternal relatives overwhelmingly on Skype – my two-year-old son asks for "Bubie in the computer" and furtively looks around the back of the screen to see where she is hiding. My sister, apart from my father at a similar age, needed coaxing into his arms when they were finally reunited because she did not recognize him. These connections remind me, in Youngs's words, of the ways that "[t]he Internet is characterized by its cross-boundary nature, and … goes *beyond* the (physical and geographical) world we have previously inhabited" (188).

If the speed and embodiment with which we now communicate has engendered real-time dialogue for transnational family in previously unthinkable ways, can we consider our current communication more authentic? Unimpaired by the lugubrious time lag of airmail letters, or the anxiety-provoking haste of timed long-distance calls, free to view one another's facial expressions and gesticulations, are we now freer to create more authentic relationships? Although it is tempting to celebrate technological innovation as a site of authentic familial communication, I think it would be incorrect to suggest that technology has led to greater authenticity. We may choose our words less carefully now that we can spend them more freely, but ultimately, we are still bouncing off of one another and creating a shared version of events. Relaxed on the phone now, my mother argues with her sisters more often, and at length. After each visit I enthusiastically email my cousins before thousands of other emails clog my inbox and I become too busy to reach out. Both speed and embodiment have shifted our capacity for dialogue, have created a dialogic view of family unrecognizable from the slow transfer of written words which characterized transnational kinship in the very recent past; while this new dialogic context is deeply altered, however, it may not transcend the difficulty of maintaining intimacy over great distance; it may foster different connections but not necessarily connections that are closer or more honest than those which came before.

Language, Grief, and Connection

The horrendous busyness of contemporary networked life is certainly responsible for some of the disconnect between families, but transnational familial communication, like other global connections, also has to transcend language barriers. My closest cousins do not read or write English and, despite a fairly competent command of spoken Hebrew, I cannot read the emails they send me. Indeed, to read or write in Hebrew would require both familial interpretation (another potential source of both dialogue and misinformation) and the reformatting of my computer as well as the installation of particular software. As a result, we are unable to use written communication as a means of reaching out to one another. This includes email, but also "The Facebook" (as they adorably refer to it) and Twitter, all technologies that we rely on as means of connecting with our Canadian friends and family.

As a result, the basis for connection is diminished; not only do we speak and write in different languages, we have different points of reference. In this, our virtual lives parallel our lived realities. Can I relate to my cousins' mandatory army service without resorting to ambivalence, or worse, repulsion? When I try to describe my life here, they cannot get past the snow: "Tell us more about how much it snows. How do you leave the house? What about the children?" Our lives are exceedingly different. Our networked lives reflect and sometimes amplify those differences, and as a result, there are barriers to our capacity to communicate that go beyond simple linguistic limitations.

On the phone, on Skype, these linguistic limitations are eased, sometimes removed. We are able to summon the intimacy of our embodied time together. Yet even in these formats, language is difficult. My older relatives are increasingly reverting to Arabic, the language of their childhood, a language that I cannot speak and can barely understand. My parents, by contrast, have created an "interlanguage" of Hebrew, Arabic, and English that baffles their siblings. There are gaps and long silences, confusions and mistakes. The precision that I have come to rely on in my written and spoken communication is absent, replaced instead with groping, vague sentences. We cannot access our deepest thoughts and feelings together, so we tell each other how much we love one another; we convey the depth of our missing. The emotion is strong, but even lightning-fast communication cannot adequately convey it.

Recently, my 77-year-old father came over. "Do you know about 'skep'?" he asked. After some questioning, we learned that he was interested in Skype and, specifically, in reaching out to his ailing aunt, a woman who helped raise him; with no children of her own, she dotes upon my father and his descendants.

Dodda Samra is my oldest living relative. At age 95, she lives at home with a Filipina migrant caregiver, a woman displaced from her own home by economic upheaval and globalization. Chris, Dodda Samra's caregiver, speaks English as her second language, rendering her communication with my great-aunt, who communicates in Arabic with little bits of Hebrew thrown in, quite complicated. Parted from her own family, Chris reminds me of Hegde's contention that "[g]lobal flows of media technologies, migration, and the unfettered mobility of capital rework old logics of domination in new global forms" (1).

Chris communicates with her children and parents in the Philippines largely through Skype and cell phone, and she has begun encouraging my father to reach out to Dodda Samra through these technologies.

Chris and Dodda Samra live in my great-aunt's apartment, the unit we call "the apartment that time forgot." In a country characterized by dizzying technological and social leaps between my visits, I could count on Dodda Samra's apartment to remain constant, right down to the uncomfortably hard sofa, the watery lemon soda, and the old-school rotary phone. Now Chris had set up her laptop and cell phone next to the rotary dial and was encouraging us to make visual contact with Dodda Samra. My father and Chris yelled at each other in heavily accented English while in the background, my 10-year-old son enthusiastically tried to explain Skype to his grandfather. After much conversation, including an interesting discussion about technology, commerce, and culture, we set a time and waited. Dodda Samra wasn't in the apartment time forgot. Instead, Chris took her computer to the hospital, where she would be staying overnight with my great-aunt. Suddenly, we were at the side of her hospital bed, looking into her beautiful lined face. My father kept desperately yelling at the screen "We love you!" in Hebrew and Arabic, unsure of which she would understand, given her age and decline. Dodda just wept quietly between the beeps and flashes of the monitors that counted out the remainder of her days. She raised her hand to the screen, unable to communicate at all beyond the grief in her eyes and the waving of her hand.

There was something breathtakingly lovely in connecting with Dodda Samra. It was surreal to see her outside of her environment, surrounded by modern medical technology and viewed over a wireless connection, when her context has been stuck in one place since 1951. I cannot imagine what this woman, born in 1918, understood from our smiling, stricken faces on a screen, from our frantic cries of love. While it was a great gift to see her, perhaps for the last time, it was not an entirely comfortable connection. As with other technologies, our capacity to communicate results in a drive to do so at any cost. Was it better to interrupt Dodda Samra's convalescence with our connection? My heart soared to see her, but broke to see her so diminished, and as we tried all our languages, all our physical and linguistic flourishes, to connect, my heart sank. We could not connect. We could only agitate her, confuse her, and in the process upset ourselves. The connection ended with all of us in tears. Was this better than continuing the narrative of understanding Dodda Samra in all her 1951 glory? Was this twenty-first-century intervention productive or merely brutal?

Archive

The interaction with Chris and Dodda Samra was simultaneously beautiful and brutal, but when it was over, it was gone. In this regard, this moment differed considerably from much of the communication that has defined my family over the last 50 years. Historically, our primary communication was through tangible means. The shoebox full of letters will exist for me to share with my children; perhaps one day we will translate them together and sit with a record of my parents' love and lives. By contrast, our interaction with Dodda Samra was fleeting and intangible, both distinct from and reminiscent of ways we had "talked" with her in the past.

I am a newlywed, sitting in the apartment that time forgot. Dodda Samra, having already served the watery juice and cut me an apple that I didn't want, has begun reminiscing. She discusses her longing for life in Iraq and her own interesting exodus: leaving before the mass harried flight of my parents' own departure from Iraq to Israel, she and her husband went to Tehran instead. Departing at leisure, they maintained their financial status and purchased household belongings that eventually arrived in Israel by boat, the same items that I now survey as I sit on her uncomfortable sofa. Grateful for the fresh audience of my new spouse, she presents little treasures like a magpie – part of her wedding dress, the photo of my long-dead grandfather, old-fashioned jewellery. Suddenly she claps her hands and brings out several old reel-to-reel recordings and explains that these are from my father, sent from Canada nearly 30 years prior. She presses them upon us and compels us to figure out a way to draw out the memories from this extinct technology.

Upon our return, my partner tinkers with the reels at the university media lab and manages to extract ghosts – my father, strangely younger and high-pitched; my sister's baby voice, singing in English with a heavy Hebrew accent. We convert the recordings to audio cassettes and, on our next trip, buy Dodda Samra a tape player so she can listen. She beams with delight, praises our brilliance. Now, 15 years later, we can scarcely find a player on which to play the cassette tape, finding ourselves again held in the grip of obsolescence.

As the technology has evolved to allow my family faster and more plentiful connection, strangely, the archive has diminished. Through each shift in technology, our capacity to access the stories of the past is lessened, such that the quaint and old-fashioned handwritten letters between my parents become our most effective representation of the past. By contrast, the recordings have become as inaccessible as the large box of photographic negatives from that long-ago newlywed trip. While both expense and effort may be applied to salvage these records, they are nowhere near as accessible as the thin paper accounts of my parents' longing for one another.

There is a further challenge as, in some regards, our archive has become so voluminous that it is impossible to catalogue, let alone read and understand. It is difficult for me to imagine my children rifling through my computer and the tens of thousands of emails sent between their father and myself with the same romantic fascination with which I approached that shoebox so many years ago. Likewise, our photographic records display an overwhelming excess. There is exactly one photograph of my grandfather in the world. We have all seen it, all have a copy displayed somewhere. In the face of the thousands of images that chronicle my present-day life and my recent history, I rarely know where to look, what to find. The images sleep in my hard drive under a heavy blanket of their own excess; I haven't printed an image in years. Part of me delights in the ways that the story has become convoluted and complex – the dizzying array of connections and lightning-fast communication have allowed us to construct an evolving family narrative that resists easy categorization, and I revel in the ways that "stable truths appear to have become unstable truces" (Iedema and Caldas-Coulthard 1). Yet part of me wonders if our capacity to trap everything has resulted in us effectively trapping nothing, has resulted in an incapacity to access anything at all. In the static of the tens of thousands of images of my children, will their grandchildren ever know that they, like their grandfather, stood in front of Niagara Falls? Does it matter?

The excess does not merely impede our capacity to access archive; it may limit our ability to reach out at all. In the midst of the reams of words sent to me every day, I feel suffocated, swamped by communication. In order to write this chapter, I needed to invest in technology that cut my Internet connection (a stunning example of a first-world problem, I'll admit) because the never-ending deluge of words, instructions, requests, photos, movies, and tasks can become so relentless that I have no capacity to stop, to pick my words deliberately, to communicate with thought and ease. At the same time, I revel in the mundanity of our networked lives: the capacity to text a request for more milk, to vet my child's latest choice in music video and send him an email ("No, this one has inappropriate words"), to use my voice and my body to comfort a friend who cannot make it to my sofa for the embodied support I would like to offer. My relationship with technology, with networks, is thus ambivalent and convoluted.

Yet through the excess, I maintain a commitment to ensure that the family has a story, that we have a means of conveying where we are and where we come from. Perhaps this commitment is born of my family's history of trauma, of the requirement that we "enact the possibilities of hope through a required meeting with traumatic traces of the past" (Simon, Rosenberg, and Eppert 5). In many respects, my relationship with communication technologies is thus much like my relationship with family. When I go to visit, my giant family swarms me with their love and attention, fills me with food and words until I am literally and figuratively stuffed. I swoon at their excess, wish there were a way to disable my network with them briefly to catch my breath, labour under the weight of their heavy love; at the same time, I feel the "possibilities of hope" bound in our ongoing connections. In this respect, as in many others, my family history is truly a history of both communication and transnationality, the shifting of our gushes of words through different media, across time and space, navigating identity, language, and grief through the archive of our love and longing.

In loving memory of Samra Shahrabani, 1918–2014, z"l.

WORKS CITED

Alarcón, Norma, Caren Kaplan, and Minoo Moallem. "Introduction: Between Woman and Nation." *Between Woman and Nation: Nationalisms, Transnational Feminisms, and the State.* Ed. Caren Kaplan, Norma Alarcón, and Minoo Moallem. Durham, NC: Duke University Press, 1999. 1–16. Print.

Butler, Judith. *Gender Trouble: Feminism and the Subversion of Identity.* New York: Routledge, 1990. Print.

Friedman, May. *Mommyblogs and the Changing Face of Motherhood.* Toronto: University of Toronto Press, 2013.Print.

Hall, Stuart. "Cultural Identity and Diaspora." *Identity, Community, Culture, Difference.* Ed. Jonathan Rutherford. London: Lawrence and Wishart, 1990. 222–37. Print.

Hegde, Radha S., ed. *Circuits of Visibility: Gender and Transnational Media Cultures.* New York: NYU Press, 2011. Print.

Iedema, Rick, and Carmen Rosa Caldas-Coulthard. "Introduction." *Identity Trouble: Critical Discourse and Contested Identities*. Ed. Carmen Rosa Caldas-Coulthard and Rick Iedema. New York: Palgrave Macmillan, 2008. 1–15. Print.

Kadar, Marlene, ed. *Essays on Life Writing: From Genre to Critical Practice*. Toronto: University of Toronto Press, 1992. Print.

Lemke, Jay. "Identity, Development, and Desire: Critical Questions." *Identity Trouble: Critical Discourse and Contested Identities*. Ed. Carmen Rosa Caldas-Coulthard and Rick Iedema. New York: Palgrave Macmillan, 2008. 17–42. Print.

Mohanty, Chandra Talpade. *Feminism Without Borders: Decolonizing Theory, Practicing Solidarity*. Durham, NC: Duke University Press, 2003. Print.

Shohat, Ella. *Taboo Memories, Diasporic Voices*. Durham, NC: Duke University Press, 2006. Print.

Shohat, Ella, and Robert Stam, eds. *Multiculturalism, Postcoloniality and Transnational Media*. New Brunswick, NJ: Rutgers University Press, 2003. Print.

Simon, Roger I., Sharon Rosenberg, and Claudia Eppert. *Between Hope and Despair: Pedagogy and the Remembrance of Historical Trauma*. Oxford: Rowman and Littlefield, 2000. Print.

Youngs, Gillian. "Cyberspace: The New Feminist Frontier?" *Women and Media: International Perspectives*. Ed. Karen Ross and Carolyn M. Byerly. Malden, MA: Blackwell Publishing, 2004. 185–209. Print.

9 Love Knows No Bounds: (Re)Defining Ambivalent Physical Boundaries and Kinship in the World of ICTs

ISABELLA NG

Introduction

As a 40-something Chinese woman who grew up in Hong Kong, was educated, and has worked in different places around the world, I have witnessed and been affected by the staggering growth and advancement in new media. I have experienced the transformation that new media technologies have made in my work, but also in my personal life, as when members of my immediate family migrated to different parts of the world, facing the uncertainties of the handover to China and the changing political and socio-economic dynamics in the ex-colony.

Increasing mobility and opportunities to move around for work and study at the beginning of the 1980s in Hong Kong also contributed to the increasing need for and use of telecommunications and eventually quick media to connect with relatives. Like the Chinese professionals in London described by Kang and the upper-class Europeans examined by Andreotti, Le Galès, and Moreno Fuentes, the idea of connecting with family or the local community to retain a sense of belonging and rootedness while not physically present, whether away for good or constantly on the move (a connection enabled by quick media in the last two decades), has created tension between feelings of foreignness and rootedness, redefining intimacy and closeness, (re)creating an imagined family to which one can connect or relate through quick media.

There has been growing interest in the effect of Internet Communication Technologies (ICTs) on transnational families, especially among low-income families (Madianou; Madianou and Miller; Parreñas). Studies by Chen and Kang have attempted to understand how male and female Chinese professionals and female entrepreneurs use ICTs to communicate with their families and make use of modern technological prowess in the transnational context. This chapter takes up this interest by looking at the impact of information and communication technologies and how they have transformed my kinship and nurturing experiences as a Chinese woman.

Drawing from my experience of a diasporic Chinese family in Australia and Scotland at the time when Hong Kong was also experiencing the angst of handover to China, I trace how the advancement of ICTs has facilitated nurturing and caring processes among family members. In particular, I will explore the ways that instant messaging and instant video communication have (re)defined physical boundaries to the point that one may not even feel separated from one's family. This chapter considers the ambiguous nature of territorialization and the ways that physical boundaries have been (re)defined by new media and technologies. It will also look at how the development of ICTs is affecting transnational family relationships and how the concepts of citizenship and nation-states are blurred by the use of new media and technologies. Finally, I will consider how identity is transformed and challenged when we are so far apart, yet so close to each other. Able to maintain our family ties and our customs through these new technologies, we are at the same time challenged by our physical surroundings. We are constantly (re)orientated and our identity is consistently (re)negotiated through the use of ICTs. This is certainly true for my personal identity narrative.

My Life as a Chinese Woman

There are many discursive ideas about what a Chinese girl is like: perhaps one of the tragic female characters in Jung Chang's *Wild Swan* or the older generation of women portrayed in Amy Tan's *The Joy Luck Club*. Often, Chinese women are thought of as domestic; bearing their suffering with silent pride; quiet, nice, obedient. In short, Chinese women are popularly constructed as the opposite of the roaring American women of the 1970s women's liberation movement.

To my friends' and classmates' (and, later on, colleagues') disappointment, I possessed none of the expected qualities of a stereotypical Chinese girl. On the contrary, I was raised to be independent, to be outspoken, and, to my parents' pride, to be adventurous as well as audacious. I was sent to camps and orienteering when I was very young.

I was not encouraged to help with any housework, which was quite unusual in a Chinese household. My parents wanted me to focus on my schoolwork and to excel. In short, they wanted me to be able to find a good job and stand on my own feet, not to depend on a man. They also made sure that I knew I was as good as my brothers and maintained that my girlhood should not make me feel that there was anything I could not do. Most importantly of all, I was encouraged to pursue higher education to whatever level I chose.

Ironically, being taught to be independent in the Chinese context, at least in my family, did not mean I could be independent from my family. I grew up with five siblings. It was pretty normal to have five siblings in Hong Kong in the 1970s. But then my siblings got married one by one, and the clan expanded with nine more little ones who one by one came to stay in my house to be looked after by my mother. As a result, I spent a lot of time helping my nieces and nephews, including helping them with their homework. My family is, by all accounts, very tightly knit, including ties with extended family. Growing up and well into adulthood, I saw my siblings at least three times a week when they came over for dinner. Sundays were reserved for family gatherings. Anything happening in one branch of the family was quickly known by other family members. The experience of being part of this family taught me that everybody's business is my business. We were expected to care about our siblings and younger family members. Even though I am the youngest sibling, my job is to care for and nurture my nieces and nephews, the third generation of my family tree. I remember that, when I was still in primary school, I had to accompany my eldest niece to nursery because she cried whenever she went there. I had to stay there for the whole morning and only went to school myself at noon. I had to help them to do their homework, and as the number of children in my household increased, I had to call my high school friends for help with this. When I went to my friends' homes, I usually took my nieces and nephews with me, thus allowing my friends to get to know them. While my family did not expect me to perform household chores, I still had to contribute as a young person, using my literacy and helping with child care.

Of course, growing up in a big Chinese family has both good and bad sides. The good thing about having lots of siblings and, in turn, many nieces and nephews, is that it is always very hectic, with lots of traffic and lots of fun. My siblings' children grew up in my home. Since my nieces and nephews were only a few years younger than me, we were as close as siblings ourselves. And since my sisters and brothers were all much older than me, they treated me like their own child. The difficulty came from the fact that since everybody's business was my business, I was expected to make sure the young ones were all right. Any mishap meant that I had to jump to their rescue: little Donna's Chinese test; little Billy could not do his maths; little Jason was still not home from school; little Jeffrey had a hundred sums from school and would be having a history test the day after. There was always something that required my presence, often related to academic help, academic decision-making, and pastoral care.

Out of Sight, Out of Mind? Chineseness Under Threat

After spending my high school and undergraduate years in Hong Kong, I went to London to do a master's degree and then worked for an international news organization, which took me around the globe. My boss and mentor, an American woman who had become a correspondent in the 1970s when the women's liberation movement was at its height, was someone I aspired to emulate. She exemplified the words of Virginia Woolf: "As a woman I have no country. As a woman I want no country. As a woman my country is the whole world" (quoted in Kaplan 137).

At the age of 27, I was experiencing the fun, the adventure, and the glamour of being a correspondent and a cosmopolitan woman. Though I was based in Hong Kong at the time, the budget and opportunity offered by my news agency allowed me myriad chances to travel around the world to work on stories and experience the adventures of being a journalist. I wanted to have a life of my own with the kind of freedom my boss enjoyed, to meet interesting people and have my own career. It seemed at that time that I was getting nearer to my childhood dream, the dream, fostered by my parents, of an independent life. I was very much my own being: independent, assertive, and always on the go. I was still close to my family as I still lived with them when I was not travelling. But my travel schedule meant I spent much less time at home.

I was happy to be single, as marriage was something very peripheral for me. Yet as my childhood friends got married one by one because the clock was "ticking," I was struggling to fulfil both of my desires – to conform to "Chinese" values and yet to be myself. I found myself trapped in my Chinese body and by the expectations of my Chinese peers: that I get married at a marriageable age (i.e., between my mid-20s and early 30s) and have babies before 40; at the same time, I felt a longing to be my own self – to be independent and able to do what I wanted. Shohat's idea of a "transnational imaginary" and other transnational feminist writings by Inderpal Grewal and Caren Kaplan struck a chord with me in my own situation. While Shohat emphasizes the "hybrid nature of communities in a world characterized by the movement of people, goods, and images" (quoted in Križ and Manandhar 222), I had already found my own self to be hybrid and conflicting, constantly experiencing the tension between my Chinese ethnicity and both my unorthodox upbringing and the feminist education I was continuously exposed to in different settings.

I witnessed how such a hybridity could be created in a community by people leaving and returning, bringing ideas, goods, and images to and from their original culture. In England, where I studied, I experienced some of the gifts and challenges of multiculturalism. At university, I had classmates from all over Europe and the United States, and we exchanged ideas and thoughts during class and after class over drinks and dinner. In the dormitory where I lived, there were people from all over the world: England, Thailand, Malaysia, Greece, Singapore, Germany, Sweden, Korea, Botswana, and Zimbabwe, to name just a few. They brought information and knowledge about their home countries. We experienced our different cultures together. This was especially so when we cooked together. We shared our different cuisines and our stories. I learnt about their cultures and took up cultural moments and foods as part of my own. My roommates did the same. For example, even now, I make couscous with African chicken when I have time. My German friend still uses the black bean sauce I introduced her to when cooking in England. During my time working for an international magazine, I also witnessed how diversity and hybridity were percolating into every corner of the community and creating a transnational imaginary. My workplace, situated in the middle of a very small neighborhood downtown, stood unmatched with its shiny and modern exterior. Five stories of the 40-story building belonged to a gargantuan organization packed with workers from the United States, China, India, the United Kingdom, and Australia. There, we learnt one another's accents and shared our own special habits acquired through our upbringing and our interpretation of the local Hong Kong culture. With information, emails, and report files flowing to and from New York, Japan, Korea, China, and the rest of the world, the daily editorial meeting allowed us to share news and thoughts on anything happening around the world, recalling the transnational imaginary as described by Shohat.

I am no longer defined only by where I am from. I am the combination of what I am, what I have been, and what I will be. My Chineseness is no longer adequate to define me. This is mainly because, on top of my experience of different worlds, learning from each, I am also situated in this transnational imaginary, which is further accentuated by the rapid development in ICTs.

The globe is becoming more connected because of the drastic increase in the movement of people, goods, and information, brought about especially by ICTs. Friedman and Schultermandl suggest,

> In this context, Stuart Hall's (1996, 608) concept of "identification" as a conglomeration of the ways in which we position ourselves and in which we are positioned by others gains additional importance: no longer does it suffice to look into specific locations, such as within a nation-state, to determine identities. We must also look into how these locations confine, interfere with, and contradict individual projects of selfhood. (13)

It was not just my independent identity but the Chineseness of my family that was under threat when both of my sisters' families moved in 1997. While this was a common phenomenon in Hong Kong in the run-up to the transfer of sovereignty that occurred that year, we were working through ways in which we could maintain our ties. There were emails and IDD (international direct dial) phone calls. But these communication technologies were not popular among the older generation in Hong Kong. I had been using IDD phone calls and email a lot because of my work. But what about our parents, who were already in their sixties? What about my brothers, who were not very keen on phones? This was before the advent of ICQ or MSN Messenger or Facebook, let alone a freemium video voice-over and IP messaging service like Skype.

Despite our initial high expectations, things changed dramatically as the only ones who kept in touch with my two sisters were myself and my mother. We no longer had 20 cramped people in a 400-square-foot flat eating dinner together. My sister in Australia, my niece, and my mother were still talking at least three times a week; my niece and nephew would occasionally send photos to my mother. But the hectic traffic was gone. With the exception of Chinese New Year, my relatives in Australia and Scotland seemed to gradually forget the Chinese festivals, which passed without fanfare in England or Australia back in the early 1990s and early twentieth century. We tried to send moon cakes during mid-autumn festivals but Australia did not allow food as postal gifts, and my sister soon received a letter from the customs department telling her the cakes had been confiscated.

As a result of these limits to our communication, little by little, these memories of Chineseness were disappearing from the second generation. My nephews, who were then in primary school in New Zealand, and subsequently in Australia, no longer read and write Chinese. They speak Chinese with a weird accent. For example, they once asked me, "Are you dead yet?" when in fact they meant, "Have you bathed yet?" The two words "*sei2*" (dead) and "*sai2*" (washed) are not homonyms, but if one twists the accent of one word it sounds like another. It was hilarious but at the same time sad: I felt that these children, who were very dear to me, were losing part of what I am.

Despite my hectic schedule, I was still making an effort to visit the children in Australia once a year and spend at least a month with them, helping them with their homework and with their teenage angst, anger, and anxiety since, from their perspective, their parents were completely clueless and often the source of conflict.

As I was tending to my Australian family, my niece in Scotland had just finished her undergraduate studies and was seriously dating someone. She indicated to us that she was going to get married soon and that she intended to remain in Scotland. Her husband is a first-generation migrant from Hong Kong and owns a Chinese takeaway shop. Since his work was in Scotland, my niece would naturally stay in Scotland and build her family there. My mother and I still maintain close contact with my niece since she was raised by my mother. My mother talks to her every other day, and my niece sends letters, cards, and videos to my mother from time to time. Since her son was born in 2004, our tie has become even closer, with my niece calling every day asking for parenting advice. She also sends pictures and videos more often than before. We are still close, but we also see that the intimacy is becoming difficult to maintain because of the distance that separates us. We need to put a lot of effort into making phone calls and sending emails and letters, cards and gifts, in addition to annual visits, to ensure that our kinship and love do not fade away, especially with the third generation. The tension of connection, however, has shifted since my family and I began to set up new media technologies at home to keep in touch, once again changing our roles in one another's lives.

(Re)Visiting Kinship Ties and Bonding – with the Evolution of Quick Media

In the globalized era of expanding mobility, transnational families are an increasingly common phenomenon. Quick media, which facilitate a faster, more instantaneous means of communicating with people, have altered the way families connect with each other and taken the definition of kinship to a whole new level. In a number of studies on the relationship between kin relations and quick media, it has been found that the idea of defining kin and relating to kin has changed. Kang found that quick media have altered the way that Chinese professionals relate to their aged parents, with the traditional feminine role of care and intimacy now increasingly redistributed to male family members as quick media become an important means for transnational family communication.

There is a body of literature demonstrating that quick media are technologies which alter social relations of transnational families through those families' mobility, especially how transnational families are parenting. Recent studies on transnational migration have examined the hybrid and negotiated process of gender roles in terms of parenting through new media technologies. Peng and Wong note that recent research (Fresnoza-Flot; Madianou and Miller; Kang; Wilding) argues that the "increasing popularity of mobile phones and the Internet, paired with the declining costs of telecommunication, have enabled transnational mothers to actively perform their maternal duties across national borders" (492). They further point out that by incorporating breadwinning into motherhood and highlighting mothers' economic contribution to the family and future development, transnational mothers "subvert traditional conceptions of motherhood that emphasize close mother-child interactions in an isolated setting and the dominant roles of nurturing and caring in maternal practices" (493). Hondagneu-Sotelo and Avila argue that the role of nurturing has taken a new turn, from relying on mothers' physical presence to relying on mothers giving instruction to caregivers via new media technologies. With the growing phenomenon of feminization in migration, transnational mothering has become increasingly prevalent and scholars have begun to look into the impact of this emerging phenomenon. Parreñas examined "left-behind" children in the Philippines and noted that mothers who migrated were still expected to perform maternal duties such as caring and emotional nurturing. Madianou examines the relationship between migration and a growing ambivalence towards motherhood. Looking at Filipino female workers who adopt ICT in mothering, Madianou discusses how ICT has become an important means of enabling intensive mothering at a distance. The research of Madianou shows that Filipino mothers who migrated before ICT use became prevalent found that their children did not recognize them. Madianou and Miller reveal, by contrast, that mothers with access to numerous technology platforms are able to provide their children mothering which is "more complete," as each technology is used for a different purpose. However, these mothers are finding that there is a tension in their dual identity as both breadwinner who stays abroad and caring long-distance mother. These studies focus on mother-child relationships, especially among migrants of working-class origins; few studies have focused on kinship outside parent-child relationships.

It is increasingly necessary to look at how family members relate to each other through new media technologies outside the parent-child relationship. In my case, I do not feel the tension that many Filipino mothers report in caring for their children at a distance. This may be true for several reasons. For one thing, the nature of transnational mobility is classed and gendered (Andreotti, Le Galès, and Moreno Fuentes; Chiang). Studies of middle-class or professional transnational families inevitably reveal a different pattern of mothering and kinship from those of the female migrants who work as domestic helpers. Professionals and upper-middle- or middle-class transnational families with children, generally speaking, do not have to leave their children behind. My case is further distinguished in that my kin are not my children, which ensures that I am not subject to the dominant expectation of intensive mothering. Part of my cultural legacy suggests, however, that the job of nurturing the next generation does not fall only to the parents. Grandparents and members of the extended family are also actively responsible in the child-rearing process. In a traditional Chinese family, every family member believes that it is part of his or her job to be involved in nurturing the young ones. Unlike many modern parents in Hong Kong families, where parents and their domestic workers are often the main agents for child-rearing, older generations of Hong Kong Chinese rely heavily on close relatives for child care and nurturance and, as a result, there is a strong bond of love and responsibility between me and my nieces and nephews and, more recently, my grand-nephews. This kind of love and trust is different from, but in some respects may resemble, parent-child love. For example, many years ago my eldest nephew told a leader of his church, "My mummy is my mummy; auntie is very dear and special to me. No one can replace her." The strong bond between me and my nephew has been nurtured through years of cumulative love and care via visits and, most importantly, via the continuous advances in ICTs. The nurturing of kinship via quick media has also redefined the notion of kinship, based on the choice of people with whom one wants and feels a need to connect, to continue this intimacy.

With the advent of different ICTs, from purely written communication (including instant messaging) to audio and then to video calls, my family members began to explore the different media and were ecstatic to realize that it was so easy to stay in touch. The discovery and the process were exciting and often reassuring. We now use Skype for group prayers and group chats on family issues. It is also convenient for my eldest sister, who has now returned to Hong Kong, to chat with her grandson when they miss each other.

Despite the miracle of connected technologies, I do feel ambivalence about physical boundaries. As Appadurai writes in *Modernity at Large*, new electronic media have broken down all physical barriers and created a number of new virtual groups. The boom in ICTs has facilitated the diasporic communities' ability to create "a more complicated, disjunct, hybrid sense of local subjectivity" (197). The radical development in ICTs has led to constant (re)orientation of the locale and the context; in my own case, it is especially evident when I am travelling frequently and use Skype in hotels to connect with my family. On the one hand, I enjoy the convenience and the intimacy that ICTs offer. I can talk to my relatives anytime, anywhere. On the other hand, ICTs, which allow me to traverse the physical boundaries between myself and my relatives, have also created a free zone so apparently infinite that it is overwhelming and befuddles users like me. Sometimes, when I speak with family via Skype for more than an hour, I begin to lose my sense of location. I feel disorientated and amazed at the same time because I don't know exactly where I am. It seems like I am at home with my family but at the same time I realize that we are apart. I know on a conscious level that we are in two different countries, in two different time zones. But it seems that they are actually in the same place as me. At this point, the ideas of nation-states and territory may feel less relevant when the computer or phone is switched on and one is online. ICTs subvert the physical boundaries in the real world and (re)establish the imagined community as real. The two worlds, in two different time zones, are united and yet apart. The world becomes fluid, hybrid, and unstable, and boundaries seem to be constantly negotiated when we use these different technologies. As Ella Shohat and Robert Stam point out,

> By facilitating a mediated engagement with distant people, the media "deterritorialize" the process of imagining communities. And while the media can destroy community and fashion solitude by turning spectators into atomized consumers or self-entertaining monads, they can also fashion community or alternative affiliations. (145)

The function of cinematic media in Shohat and Stam's observation resembles that of ICTs. With the booming evolution of new media technologies, ICTs have served multiple functions in our increasingly complex and mobile world. They have de-territorialized physical boundaries while re-territorializing communities and affiliations. They create opportunities for distant people to come together while potentially destroying existing communities in the "real" world. While my family and I have benefited tremendously from the rapid growth of new media technologies, we have also seen the potentially disruptive nature of the growing demand for ICTs. My students have told me that they spend more time on the Internet than in conversation with their parents over the dinner table. Parents often complain about the indifference of children towards family affairs and report being ignored. A lot of students spend most of their time in Internet forum discussions with people they may have never met, or Facebooking with friends rather than having an actual face-to-face conversation with their friends. Some of my students have told me that they find it exciting to meet girls or boys via the Internet but that they are too shy to meet them face-to-face because they do not know how to handle an actual relationship taken out of the virtual world.

Grewal and Kaplan identify new forms of Western hegemony and warn against increasing cultural homogenization, but in both Filipino mothers' and my own case, it seems that though using ICTs as a means of nurturing leads to the potential rupture of national barriers and "new articulations of spatial relations" (9), it has also helped perpetuate ethnic lineage and strengthen ethnic identity through instant, readily available, face-to-face contact. In my case, passing on Chinese culture and celebrating festivals together has become much easier now that we can just turn on the computer and see each other. In many ways, the convenience provided by different ICTs has become, to some degree, a medium of continuing our lineage, of keeping nationalities and ethnicities alive in the hearts of the next generation. In this way, nation-states and ethnicity are being reinvented through different forms of new media technology and the intersection between the local and the diasporic. While ICTs enable the perpetuation of an ethnic culture and allow for the preservation of ethnic identity, the diasporic voices from these communities have also been transported to the ethnic community in home countries. Ethnic culture is therefore (re)invented through the use of new media technologies.

Nurturing the Next Generation in the Modern, Transnational Chinese Context and in the ICT Era

The shift from pure messaging and long-distance calls to the almost inexhaustible possibilities of video-calling, 4G, WhatsApp, and WeChat has had a significant impact upon the world. The way my family communicates has also changed with the development of new technology like emails, instant messaging, Skype, Facebook, and WhatsApp. In the past 10 years, we have been able to keep in touch constantly, especially since these e-applications for communication are mostly free. In the early phase of ICTs, I mainly used long-distance calls, together with instant messenger and emails, to connect with my nephews in Australia. They would email me their essays and I would talk to them on the phone while looking at their essays on my computer. Supporting my family through online connection has become such a constant part of my life that I once was able to do it even though I had taken my students to a camp. My sister called me on the phone and asked if it was OK for me to look at my younger nephew's essay. I quickly went to the campsite's recreation room and turned on their computer, and there I was, copyediting my nephew's essay. Within 30 minutes, I was able to help, though I was 1,000 miles away. Despite the speedy medium though, I could still feel the distance. More optimistically, my personal space was still intact despite these communication technologies.

These days, I turn on Skype whenever I get home. We all know the time differences so whenever we want to talk to each other (which is almost every day), we first contact each other via WhatsApp to make sure we do not have other appointments. Then we turn on Skype. Very often, this ritual is more than just having a conversation. We leave

Skype on while we are doing our housework separately in different households, across different time zones. We leave Skype on for hours without talking to each other. We usually focus on conversation in the first hour, more like a traditional telephone conversation. After an hour or so, however, we spend time together more passively: sometimes I take a shower while my niece is working on her homework, other times I work on the computer while my niece cooks in the kitchen. It feels like we are in the same house, but we also know that we are in two different places, across two different cultures. We are together but apart.

Skype has become a platform for group chat, too, when my sister in Australia and my niece in Scotland want to talk to me simultaneously. We do group prayers together or discuss what's happening to other siblings and their families in Hong Kong and beyond. These days, whenever there is an important Chinese festival, my niece and I turn on Skype and eat together (seven to eight hours' difference means that I have my dinner while they eat lunch). My niece and her son are more aware of the festivals and the fun and fanfare during the festivals are equally shared because of the disjuncture and the liberation of space that Skype engenders.

Pedagogically speaking, new technology allows me to continue helping my nephews and my grand-nephew with their homework beyond the time I spend in their homes each year. I can see them face-to-face, listen to their problems, correct their pronunciation, and instruct them on what to do next. Most of the time, our video-chatting is coupled with other ICTs, such as email, through which they share their work and I give my feedback.

While this simultaneous communication has given my family amazing opportunities, there are also challenges. My timetable presents a constant struggle. Along with maintaining my heavy workload in the academy, I am also committed to this surrogate parenting, in addition to travelling frequently for conferences, talks, and speeches. As Friedman and Schultermandl note, "no unified theory of transnationalism can be applied to the numerous situations in which increasing mobility and immense cultural exchanges occur" (18). While co-raising my grand-nephew with my niece and her husband, I have discovered how my transnational background and theirs converge and how motherhood, parenting, and guardianship are (re)defined by increasing mobility and cultural exchange. My Hong Kong upbringing affects the way I perceive the importance of academic excellence while, at the same time, having lived overseas for a long time propels me to encourage my grand-nephew to explore ideas and things that are discouraged in the Hong Kong education system, notorious for its spoon-feeding approach and a tacit acceptance of social privilege. Likewise, my niece brings what she has learnt from her education degree in Scotland into her teaching of her child. The child's father, a first-generation Chinese migrant, feeds him traditional food and teaches traditional ways of eating and celebrating different Chinese festivals; he also has a traditional way of fathering coupled with a heightened intimacy with his son (seen through morning rides to school and rides to swimming lessons). We see how this child has become a by-product of his transnational upbringing, with the help of new media technologies.

Through our intense nurturing of children and young people, my niece, my sisters, and I are also becoming by-products of new communication strategies. We have created our own "virtual neighbourhoods" (Appadurai, 198). Coined by Arjun Appadurai, this term refers to clusters of people who have access to international computer networks and who form groups to trade information and build links that affect many areas of life. My sisters, my niece, and I now use Facebook, Skype, and email to update each other on our lives, share information and news on our locales, and exchange thoughts and our perspectives on families. Our situation reminds me of the movie *Hanging Up*, in which the three sisters – played by Diane Keaton, Meg Ryan, and Lisa Kudrow – are always on the phone updating each other about what has been going on in their lives and that of their curmudgeonly father. The only difference is that instead of relying on the phone, we are now using ICTs.

As Kaplan writes,

A transnational feminist politics of location in the best sense of these terms refers us to the model of coalition or, to borrow a term from Edward Said, to affiliation. As a practice of affiliation, a politics of location identifies the grounds for historically specific differences and similarities between women in diverse and asymmetrical relations, creating alternative histories, identities, and possibilities for alliances. (139)

My sisters, my niece, and I – growing up in the same family, belonging to different generations, and undergoing different life experiences – are trying to understand each other, share, debate, discuss, and, at times, challenge each other's views. Through chatting on issues ranging from family, children, and love to religion and current affairs, we are creating an "alliance" based not only on blood and lineage but also on sharing different experiences as women, to make us feel united even though we are apart. All these interactions are made possible through ICTs. The disjuncture, the rupture created by new media technologies, creates an intimacy that transcends all physical barriers, even though we are consciously aware of our differences in time zone and locale. Our differences in experiences and histories have ironically united us all, as women who are unique and distinct in our own ways.

My autoethnographic account of altered kinship through the development of quick media has shown the myriad possibilities of quick media in transcending physical boundaries and reconfiguring one's identity in relation to surrounding people, whether they are family members, extended family, or friends. Through the evolution and swift development of quick media technologies, kinship has been redefined to include not simply a link that relates physically to a locale and lineage but also one based on choice. My autoethnography shows that while I actively pursue connection with my relatives in Australia and Scotland, my two brothers have chosen to be less involved and attached to my family overseas, even though they are also media-savvy. The "imagined family," which is connected and sustained through quick media, is a community based on affiliations, something shared in common, but also, more importantly, it is a community based on choice. It (re)defines the concept of kinship, which in conventional wisdom generally means a familial relationship based on lineage and marriage. Kinship in the quick media era means the affinity we create and sustain beyond natural connections through blood and lineage. It is a feeling of familiarity, resemblance, and camaraderie based on the willingness and the choice to (re)configure such a community.

This chapter examines the impact ICTs have had on me and on my family before and after migrating to different parts of the world. Through engaging autoethnographically with my own family experience, I delineate the process of using ICTs to maintain kinship and family ties. By tracing the development of ICTs through my 20 years of experience, I reflect on how this has transformed my kinship and nurturing experience as a Chinese woman in an extended family. I also explore how my life experience has (re)shaped the way I connect with my family and nurture the younger generation with the help of new media technologies. I also discuss how ICTs have created another form of maternal lineage and matrilineal alliance through different generations across different time zones; how women of the same origin but with different life histories and life experiences (re)connect, (re)negotiate, and (re)discover each other. We pass our knowledge; we share our different struggles in our lives; we become part of the globalized community, united yet different, all made possible through the use of ICTs.

WORKS CITED

Andreotti, Alberta, Patrick Le Galès, and Francisco Javier Moreno Fuentes. "Transnational Mobility and Rootedness: The Upper Middle Classes in European Cities." *Global Networks* 13.1 (2013): 41–59. Print.

Appadurai, Arjun. *Modernity at Large: Cultural Dimensions of Globalization.* Minneapolis: University of Minnesota Press, 1996. Print.

Chen, Wenhong. "So, Where Are the Women in Transnational Entrepreneurship?" Second International Symposium on Chinese Women and Their Cybernetwork, University of Hong Kong. 2004. Conference Presentation. Print.

Chiang, Nora Lan-Hung. "'Astronaut Families': Transnational Lives of Middle-Class Taiwanese Married Women in Canada." *Social and Cultural Geography* 9.5 (2008): 505–18. Print.

Fresnoza-Flot, Asuncion. "Migration Status and Transnational Mothering: The Case of Filipino Migrants in France." *Global Networks* 9.2 (2009): 252–70. Print.

Friedman, May, and Silvia Schultermandl. "Introduction." *Growing Up Transnational: Identity and Kinship in a Global Era.* Ed. May Friedman and Silvia Schultermandl. Toronto: University of Toronto Press, 2011. 3–18. Print.

Grewal, Inderpal, and Caren Kaplan. "Introduction: Transnational Feminist Practices and Questions of Postmodernity." *Scattered Hegemonies: Postmodernity and Transnational Feminist Practices.* Ed. Inderpal Grewal and Caren Kaplan. Minneapolis: University of Minnesota Press, 1994. 1–33. Print.

Hondagneu-Sotelo, Pierette, and Ernestine Avila. "'I'm Here, but I'm There: The Meanings of Latina Transnational Motherhood." *Gender and Society* 11.5 (1997): 548–71. Print.

Kang, Tingyu. "Gendered Media, Changing Intimacy: Internet-Mediated Transnational Communication in the Family Sphere." *Media, Culture and Society* 34.2 (2012): 146–61. Print.

Kaplan, Caren. "The Politics of Location as Transnational Feminist Practice." *Scattered Hegemonies: Postmodernity and Transnational Feminist Practices.* Ed. Inderpal Grewal and Caren Kaplan. Minneapolis: University of Minnesota, 1994. 137–52. Print.

Križ, Katrin, and Uday Manandhar. "Tug of War: The Gender Dynamics of Parenting in a Bi/Transnational Family." *Growing Up Transnational: Identity and Kinship in a Global Era.* Ed. May Friedman and Silvia Schultermandl. Toronto: University of Toronto Press, 2011. 222–32. Print.

Madianou, Mirca. "Migration and the Accentuated Ambivalence of Motherhood: The Role of ICTs in Filipino Transnational Families." *Global Networks* 12.3 (2012): 277–95. Print.

Madianou, Mirca, and Daniel Miller. *Migration and New Media: Transitional Families and Polymedia.* London: Routledge, 2012. Print.

Ong, Aiwha. *Flexible Citizenship: The Cultural Logics of Transnationality.* Durham, NC: Duke University Press, 1999. Print.

Parreñas, Rhacel. "Long Distance Intimacy: Class, Gender and Intergenerational Relations between Mothers and Children in Filipino Transnational Families." *Global Networks* 5.4 (2005): 317–36. Print.

Pe-Pua, Rogelia, Colleen Mitchell, Stephen Castles, and Robyn Iredale. "Astronaut Families and Parachute Children: Hong Kong Immigrants in Australia." *The Last Half Century of Chinese Overseas.* Ed. Elizabeth Sinn. Hong Kong: University of Hong Kong Press, 1998. 279–97. Print.

Peng, Yinni, and Odalia M.H. Wong. "Diversified Transnational Mothering via Telecommunication: Intensive, Collaborative, and Passive." *Gender and Society* 27.4 (2013): 491–513. Print.

Shohat, Ella., ed. *Talking Visions: Multicultural Feminisms in a Transnational Age.* New York: New Museum of Contemporary Art; Cambridge, MA: MIT, 1998. Print.

Shohat, Ella, and Robert Stam. "From the Imperial Family to the Transnational Imaginary: Media Spectatorship in the Age of Globalization." *Global Local: Cultural Production and the Transnational Imaginary.* Durham, NC: Duke University Press, 1996. 145–70. Print.

Waters, Johanna. "Flexible Families? 'Astronaut' Households and the Experiences of Lone Mothers in Vancouver." *Social and Cultural Geography* 3.2 (2002): 117–24. Print.

Wilding, Raelene. "'Virtual' Intimacies? Families Communicating Across Transnational Contexts." *Global Networks* 6.2 (2006): 125–42. Print.

10 The Internet Is Not a River: Space, Movement, and Relationality in a Wired World

Small distance is not already nearness. Great distance is not yet farness.
Martin Heidegger

In this chapter, I approach the notion of "kinship" in the broadest possible anthropological sense and focus on the *webs of social relationships* without which such institutions as language, meaning, identities, "societies," or even private thoughts and experiences wouldn't be possible (Tomasello). As many have argued, indeed, our species' uniquely social propensity is the linchpin that holds and defines the very experience of what it is like to be human (Sterelny). What happens to human sociality, then, as it becomes increasingly mediated online? What happens to couples, nuclear families, and caregiver/children relationships as they lose spatial grounding and become increasingly distorted and tele-mediated in the imagined communities of the digital age? This chapter will attempt to answer this question by drawing on ethnographic, materialist, and ecological approaches.

As an anthropologically trained, phenomenologically inclined theorist of sociality, I seek to avoid the mistake of presenting social worlds that exist only as cognitive schemas held together in a vacuum of free-floating ideas. That we live in webs of meaning which we ourselves have spun, as the old Weberian adage goes, is taken as self-evident by most social scientists; just how those webs are spun, distributed, stabilized, and negotiated, however, is a much harder question. Here, following others in the study of material culture, I propose to think of the distribution of ideas and the negotiations of human relations as partially dependent on materialization (Keane; Olsen) – that is, on the things and objects of our own making to which we assign meaning and function and to which our lives are constantly adapting. In order to ask how human relationships are shaped and transformed by the cheap, easily accessible, and omnipresent *tools* of quick media, I also need to ask how we continually form relationships *with* those very tools and the world at large. I should begin, then, with a brief discussion of tool ontology.

"All distances in time and space are shrivelling," proclaimed Martin Heidegger in the opening words of his 1949 Bremen lectures (Heidegger, *Bremer* 5). Radio waves, air travel, "stop-action photography": these were so many readily available technologies that carried the promise of bringing information, people, and things closer together. We were still a long way – or just a logical conclusion – away from the Internet. But, as a thinker who was more interested in pointing to unexamined questions of *absence* behind facile celebrations of *presence*, Heidegger was unimpressed with technology (Harman, "Technology"): "The hasty removal of all distance brings no nearness," he famously proposed, "for nearness does not consist in a small amount of distance" (Heidegger, *Bremer* 5).

The philosopher of *Being and Time* was noted for his early analyses of tools and his later reflections on newer technologies. As the contemporary critic Graham Harman points out, however, Heidegger was more preoccupied with a general *ontology* of objects than a mere theory of technology (Harman, "Technology"). Questions of nearness and distance – or presence and absence – against a broader interrogation of *what is real* are not easily resolved in Heidegger. In contrast with the early works of his teacher Edmund Husserl, he was steering away from the notion that phenomena can only be apprehended in the way they appear to human consciousness (Harman, *Tool-being*; Harman, "Technology"). If these latter views can be traced back to Kant's idealism – or even Plato's before that – the Husserlian tradition of phenomenology strongly solidified notions of reality as a project limited to human cognition and perception: the by now commonsensical idea of a psychologically (or later socially) constructed world. Heidegger was more interested in things in themselves or, more modestly, in the particular *relationships* we can form with things and the non-human environment.

A recent philosophical turn in the social sciences – a movement sometimes called speculative realism, speculative materialism (Harman,

"Technology"; Meillassoux; Stengers; Serres; Latour), or post-humanism (Wolfe; Pickering) – is now seeking to examine human questions against the background of so-called object-oriented ontologies. But, as media theorist Jussi Parikka argues, these post-humanist approaches shouldn't be understood as a renewed enthusiasm for the "hard" sciences, and much less as a claim that we are no longer human. "Posthumanist theory," offers Parikka, "is less about what comes after the human than what constitutes the non-human forces inside and beyond the form of the human" (210).

Here, I ground my critique of technologically mediated human relationships in such an approach. At its least ambitious, this chapter interrogates the qualitative human losses brought about by quantitative increases in travel, communication, and information technology. Following Heidegger, my interest in human-object relations goes beyond the particular relationships we have formed with specific human-made technologies and asks much broader ontological questions. As such, I am careful not to present a unidimensional world with *one kind* of human, *one kind* of technology, and *one class* of pre-existing non-human objects – a thing sometimes called "nature" – that is either revealed to or veiled to human perception. Airplanes, the Internet, and credit cards (to name but a few) are simply objects created by humans; as dominant as they have become in the way most people relate to the world and each other, they reflect as much as they shape a set of specific world views articulated by specific humans.

As an anthropologist with philosophical inclinations, I am interested in the possibilities that are offered but also those that are *not offered* by specific human world views. What preoccupies me here is the open-ended set of possibilities about what is real and how to live in the world that is no longer permitted once most of humanity is recruited into a particular ontological regime, in this case that of the "wired" world. I should translate this claim into another kind of ontological language – one recently made famous by a group of anthropologists working on questions of cosmology among different groups of Amazonian peoples (Viveiros de Castro, *A Inconstância*; Viveiros de Castro, "Cosmological Deixis"; Viveiros de Castro, "Perspectival Anthropology"; Descola, *In the Society of Nature*; Descola, *Spears of Twilight*; Kohn; see also Wagner). Noting similarities between a wide set of human pre-colonial groups (Amazonia, Siberia, Melanesia, and Aboriginal Australia) previously described as "animist," they pointed to cosmologies which, in their myriad forms, presupposed a "spiritual unity" and "corporeal diversity" (Viveiros de Castro, "Cosmological Deixis" 470) between humans and animals. This could be contrasted, they argued, with the "objective universality of body and substance [and] subjective particularity of spirit and meaning" posited by Western world views. If Western views presuppose that there is *one* "real" nature punctuated with, and variably obfuscated or revealed by, a multiplicity of cultural views (from mono-naturalism to multiculturalism), animist views propose the exact opposite: there is *one* culture shared by all living creatures (and by extension all human groups) but multiple natures, or multi-naturalism (Viveiros de Castro, "Perspectival Anthropology").

The question I am asking in this chapter is not so much what kind of "nature" is the wired world but what are all the other kinds of natures we do not live in as a result?

In examining implications for family and kin, I pose the question of relationships in as a broad a light as possible. This collection of essays defines "kinship" at its most substantial as relationships among persons (Friedman, personal communication, 2013). To understand what happens to personal relationships when they are mediated online, I must in turn pose more questions about what a person is, what she can do, and where she can go.

To introduce the problem, I begin with stories of contemporary personhoods that increasingly come to be mediated through an online mono-nature. I speak here of "mediation" in a twofold sense: one of intersubjective co-construction in negotiating the terms of personhood among humans, and the other in a more explicitly *genetic* sense: where the very existence and form but also substance of humans comes to depend on the discursive terms of the online mono-nature. A detour through travel, jungles, and rivers is necessary to get to the subject of iPhones, emails, and tele-relationships, the broader questions of personhood, space, and possibilities that they raise, and a glimpse at other possibilities. I begin in the Amazon – albeit a very different Amazonia than that described by my colleagues in post-humanist anthropology.

At the time Silvia and May invited me to contribute to this book, I had been travelling extensively in remote areas of the Amazon with rapid, interspersed stints up and down the Americas from the sub-Arctic to the southern pampas. I had recently gone through a series of relationship losses that had been precipitated by Internet events (intercepted emails, arguments over online etiquette, etc.) and was at the same time becoming aware of how intensely reliant I was on Internet communication to maintain ties with friends, family, and loved ones. My extreme mobility was making this need more apparent, and my prolonged stints in areas of the Amazon forest and river system where Internet and telephone services were unreliable or non-existent were teaching me hard lessons about *just how dependent* I was on the Internet to remain a friend, a father, a son, a lover, a salaried professor and researcher, an expatriate citizen, and a globally informed and connected person. My first forays into the jungle several years before had been deeply traumatic. Culture (or rather "nature"), symbols, and the Internet had had a lot to do with those difficulties, but it had taken me over three years to reach that conclusion.

The first symbolic layer I had to peel off and discard was my baggage as a European white male who had been culturally prepared to confront the "jungle" as a green hell of violent chaos, fevers, and unpredictabilities. I direct readers to Werner Herzog's famous comments on the "obscenity of the jungle" (on YouTube) for an inviting glimpse into that kind of violence. The second layer seemed more ontologically modest, but politically more disabling. All my relationships were strained to the point of breakage each time I re-emerged from the jungle without having attended to my email and Facebook. My inboxes were filled with so many instances of "where ARE you???" urgent HR memos, payment notices, lawyers' summons, or grant applications with long since evaporated deadlines, plus messages from loved ones threatening or announcing the end of our relationships. Yet there were strange ontological propositions imbedded in those political messages. What did a "where are you?" mean, after all, from people who were accustomed to seeing me in the flesh no more than once or twice a year? Why did my "absence" matter so much in those moments of incapacity to reply to emails? It was as though I *ceased to exist as a person* each time I went offline, as though the binary-encoded textual information I was no longer sending across electronic vectors *was somehow being mistaken for my whole person.*

I started wondering how people maintained distance ties in the age of steamboats and handwritten letters. Or before that. I began to rationalize that the consequences of my sporadic suicides were only disastrous because we decided that they were so; because we agreed to live in a world where instant online communication had become the norm.

Then something else happened. I started feeling fuller as a person in the world. This was something I hadn't felt since I was a child, when the world was naturally mysterious, unpredictable, and familiar at the same time, and when the question of whether or not I existed would have been immediately absurd.

In the petty rules of our current techno-political game, the jungle almost got the better of me. I can't say that I am sorry it didn't. When the *Tropenkoller* fear and wonder of my first years in the Amazon began to fade, annoyance and boredom ensued. I had by then cultivated enough non-attachment to the Internet to stop worrying whether I would cease to exist in the minds and feelings of my lost ones. I was doing my best to ignore the question of whether or not I would lose my job. My research project ("borders and mobility in the Amazon") was becoming a blurred priority. After another failed attempt at successfully connecting in the flesh with a Colombian girlfriend who had been a mostly platonic Internet construct, I left Bogotá on a cold day in May and decided to surrender myself to chance possibilities in the jungle.

I spent close to a week in the triple-border region of Léticia-Tabatinga, where the Amazon River intersects with Colombia, Peru, and Brazil. It took me that long to realize I was still trapped in a mono-nature, in this instance a European, Orientalizing, magical-realist one. My field notes from those days are filled with too many short poems about too many drugs, whores, a red devil, ayahuasca, a hotel without mirrors, a river full of sticks, a vampire, an Italian who could raise the dead, and a ghost-figure called *el hombre de cuero*, or the leatherman. I had to leave.

I returned to Manaus with no plans and decided to wait until people and opportunity came my way. A week later, I was boarding a boat going back upriver towards the three borders. I was travelling with a Brazilian linguist who was considering quitting everything and fleeing to the Congo and a Colombian-born Frenchman full of nightmares who was in search of his roots. We abandoned our plan to reach Peru by way of Léticia-Tabatinga three days later, halfway upriver, upon hearing rumours that Colombia could be reached in the north through the Waupés River by branching off the Rio Negro. "Perhaps it can be done, but I doubt you will succeed" – this is what an Indigenous educator working with a liberation theology cell told us in Tefê. We accepted the challenge and found a ride on a mail plane bound for the northwestern-most, remotest part of Brazil, known as *a boca do cachorro* because of its resemblance on the map to the mouth of a dog.

It is then that I came to contemplate – or participate in – other forms of connections and movement in the universe, a gradually enfolding plane of multiple natures and possibilities.

Even in large cities like Manaus, river travel is unpredictable. Phones and Internet resources provide few clues about frequency and diversity of boat schedules. Phone numbers, websites, and companies change, or they disappear. One has to walk around ports, ask around, and wait.

Things are as unpredictable in the Upper Rio Negro. Finding a port in a town or village can take a day, a week, or forever. There are many ports; people have never heard of the port; the port is everywhere and nowhere. There is just the river. One learns a lot from sitting for a whole afternoon, a whole day, or days on end by the river while people, things, and possibilities pass by. This man or that boat can take you where you want to go. It will take three days to get there. No, 10 hours. No, that man is lying, he is not going where he says he is, do not go with him, you will never get there. No, at this time of year, only an "Indian" from upriver can take you; only they know where every rock, every whirl-pool, every trunk, and every rapid lies and moves.

Time flew, and the river flowed, and we started moving again. By then, the Waupés hypothesis had been discarded. There were too many rocks, or it would take too long. The last Brazilian Indigenous village bordered Colombia, but no one could tell us what lay ahead, not even the few villagers we could find. The nearest Colombian town to figure on any map was Mitú, some 200 kilometres past the river border in the savannah. It took a day to manage to speak to someone in Mitú over the phone, and they didn't know about routes to Brazil. Maybe there was a small plane, maybe not. We went upstream on the Rio Negro instead, hoping to enter Colombia or Venezuela and reach the Orinoco, retracing in reverse the journey taken 300 years before by Alexander Von Humboldt and Aimé Bonplan (Kehlmann). But the questions of "when" and "where to" seemed unimportant. In the moment, we had food, friends, and a boat, and that was all that mattered.

The journey upriver was long and still. The river was pitch black and branched off into a million veins and pools of black water and green for-est. The boat twisted, turned, and halted to avoid rocks and rapids. How the local man could navigate through such a labyrinth without a GPS, I will never understand. Yet it seemed easy as walking along a rocky hill; feet just know where to step. I knew in that moment that the veins of my body, my blood, and the arteries of the river were one and the same.

On the way north, we passed through borders that weren't borders and posts that didn't figure on any map. We met people who lived and travelled along those black veins, and for whom international borders mattered little. Some travelled in groups to harvest gold, but most simply lived and traded in small things like chickens, manioc, fish, and açaí. Some spoke a language halfway between Spanish and Portuguese that was neither one nor the other. Many moved around in large, loose circles from Peru to the Guyanas, up and down rivers and mountains. Some knew how to be in the right guerrilla zone at the right time to coincide with the arrival of military planes and get "evacuated" for free back to Bogotá.

We emerged through the Orinoco and back to Bogotá much later. Or maybe not. I cannot recall this journey with any linear sense of time. The mono-nature caught up with us when we landed in Puerto Ayacucho, in the north of Venezuela's Amazonas province. Freshly landed from a four-seat Cessna Hawk we had managed to charter, we were greeted with military green and were duly arrested. We were more than 300 kilometres away from Brazil and possessed no entry stamp into Venezuela. A brief detention scare later, we were escorted and deported to Colombia across the Orinoco. It was the end of the jungle and the beginning of the *llanos* plains.

Had we resisted the temptation of speed and the abstraction of air travel, we could have kept moving in absolute stillness. The mistake had been to return to official, predictable, tunnelled vectors of movement when, in the grander scheme of things, all the other vectors were still possible.

The connections between veins, rivers, blood, water, air travel, the Internet, and metaphysics only became apparent to me later. I had lived a very intense experience of fluid vectors that were pregnant with possibilities and was returning to the tunnelled vector of official, predictable planetary existence with its credit-card-dependent networks of air travel, militarized borders, and standardized identities.

A little later – a freezing winter day in Buenos Aires – I realized I had made another mistake in my otherwise non-finite escape to other spatiotemporal dimensions. Touching the "map" function of my iPhone photo album after showing Rio Negro pictures to a friend, I discovered in horror that little pins had been dropped with exact precision on each point of the Amazon where I had taken a photo. I had been at times several days from any possible 3G or telephone connections; Google maps of river settlements in those areas were grotesque approximations that bordered on fiction. But the GPS-enabled geotagging function of my phone had never ceased to work and to precisely record the coordinates of my whereabouts. So much for going off grid.

Another incident pertaining to iPhone geo-tracking added more degrees of puzzlement and emotional conflict to my questioning of

cyber-ontologies. It came in the form of a heartbreaking confession from my then nine-year-old son who lives in southern Brazil when he doesn't accompany me in Canada or on my travels. I had given him an iPod the previous year to facilitate our text and video chat sessions, and had paired his device to my iPhone account. When my son really missed me, he confessed one day, he would open the "Find My iPhone" application to check where in the world I was. "It reassures me a lot to see where you are Daddy, because sometimes I have no idea." He explained that he had seen when I was in Buenos Aires and Bogotá, in Toronto and Rhode Island, or when I was at home or at work in our small Canadian town. "Sometimes, I zoom in on our blue house [in Canada] with Google Street View, and I can see our white truck; when you were in the jungle for a long time and the application would say 'unable to find device,'[1] I was really worried. I was happy when I saw you had returned to Bogotá." I tried my best to hide my tears when my

---

1 The Find My iPhone app only works if the device is connected via Wi-Fi or a 3G data plan. It could therefore not work while I was in the jungle, whereas the geotagging was still enabled for photos (and presumably the NSA).

son made this confession. How did Internet technology contribute to my ability to be a parent? Did it make me a better or worse parent? It certainly seemed to make my absences more difficult for my son. How did he feel, for example, when he "saw" me online and got no response from me?

Back in Buenos Aires after the jungle, I was becoming furious with information technology. My five-year-old son (who lives under my care) was addicted to Netflix and couldn't sit still at a restaurant without playing with my iPhone. I would blame that habit on my own negligence, then notice that other parents were dealing with similar problems. My Colombian e-girlfriend, who was in Paris at the time, was sending angry messages about my online silence. She would ask why I would post photos and poems on Instagram but not reply to her messages. She would ask me who this or that woman was when I was tagged in a photo. I would vent my anger in other online platforms, littering my Twitter or Tumblr posts with angry neologisms about FacebookNoia, iParenting, cyber-jealousy, and e-relationships. Like attracts like, and it seemed all the people I met were spontaneously vocal about similar frustrations with technologically mediated relationships: they were compulsively deactivating and reactivating their Facebook accounts. They would devise strategies for *not* going online. Groups of friends would ban mobile phones and laptops at dinner parties, arguing that friendships and soirées went on just fine before they turned into contests for the most clever YouTube 1980s clip. The ominous check-mark function indicating when a message had been read on Facebook had recently come out, along with precise geolocation map pins indicating where messages had been sent from and read. People were going through elaborate schemes to disable or circumvent those functions, in desperate attempts to prevent the tensions that would arise if they did not reply after reading. Bars, clubs, and dance halls from the Pampas to the Rio Negro were filled with immobile, lonely people shining blue Facebook light on their blank faces with their mobile devices. Back in the Amazon, I would give talks and teach classes and ask, "How many among the audience had experienced relationship problems because of email, Facebook, mobile phones, or information technology?" The question would invariably be met with a forest of raised hands. I began collecting more stories.

A 25-year-old student in Buenos Aires has four different Facebook profile photos; she interchanges them several times each week according to her mood and prides herself on her impeccably managed profile, which she equates to a tastefully designed interior. She is deeply hurt that her older boyfriend has failed to identify the photo-mood pattern and is frequently upset by his faux pas in online public etiquette.

A young woman in Lisbon who has recently gone through a break-up changes her background image seven times in one day. A friend in Toronto comments that compulsive posting and profile changing reflect instability. He proudly claims not to have changed his profile photo in two years.

A distraught boy deactivates his profile for the third time in a month. He explains that he commits Facebook-suicide when his significant other's online silence becomes unbearable, or when he can no longer live with the distress caused by her "public" interactions with other men.

An undergraduate student confesses that she deactivates her profile when she has to meet a deadline or needs to clear her head. "Facebook just has too many temptations. I can't bear to look at it, and I can't stop at the same time."

An environmentalist from Manaus echoes this conclusion. He permanently left Facebook, Orkut, Twitter, and all social media because of the problems they were causing in his marriage. Even his Couchsurfing profile will remain inactive. "I quit social media to save my marriage: too many problems, and too many temptations."

The March 2012 issue of **Harper's Index** *cites the alarming figure of 60 per cent as the numbers of American divorces in 2012 that mentioned Facebook as a reason for separation.*

"If you look nothing like your profile picture, and if you are more comfortable online than in real life, don't bother writing!" – so goes a recurring theme found in the "about you" section of a popular online dating site.

My five-year-old son, who has dark skin and brown hair, creates a tall Mii avatar with white skin and blond hair on his Wii console. He can play games online and interact with others through this avatar: "This is me when I am a grown up; I'm going to be white when I grow up."

The list goes on.

Where did we go wrong?

Now for another detour, through science fiction this time, and an early cautionary tale on addiction to virtual selves: In *Snow Crash*, his dystopian novel set in the wreckages of post-nation-state landscapes, Neal Stephenson takes us on the tracks of cyberjunkie characters in search of a mysterious drug whose effects are said to extend from virtual reality into the physical world. The story takes place in an indefinite future where countries have been reduced to big-business enclaves and residential franchised "burbclaves." We owe our contemporary use of the word "avatar" to Stephenson's *Snow Crash*, where avatars are defined as "audiovisual bodies that people use to communicate with each other in the metaverse" (33). The metaverse, in turn, is construed as "a computer-generated universe that [*main character Hiro*]'s computer is drawing onto his goggles and pumping into his earphones. In the lingo, this imaginary place is known as the metaverse. Hiro spends a lot of time in the metaverse" (24).

Snow Crash appeared in 1992 when the recently standardized Internet protocol suite was mostly limited to a closed network of military and university computers; the network was "closed" and yet open to many possibilities.

Over the next decade, the rise of commercial Internet service providers (ISPs) would slowly (then with exponential speed and reach) bring on the World Wide Web as we know it, along with the new forms of instant communication that now regulate most aspects of our professional, social, and (as I argue here) emotional and even ontological lives. Another strange thing happened in the decade following the year 2000. The broad array of electronic communication platforms that had carried the promise of so much diversity and democracy online – electronic mail, instant messaging, Voice over Internet Protocol (VoIP), two-way interactive video calls, blogs, social networking, and online shopping, dating, and pornography, to name but a few – underwent further standardization and came to be dominated by an increasingly narrow group of corporate giants; the result is, at the time this book goes to press, a World Wide Web market largely regulated by Google, Apple, and Facebook.

What concerns me here is that, as the American computer scientist Jaron Lanier (to whom we owe the term "virtual reality") put it, working with information technology necessarily entails "engaging in social engineering" (Lanier 4). The question here is how.

The birth of sociology at the height of the industrial revolution carried the first insights into how a rapidly changing world brought about new forms of violence, disorientation, and suffering. Most famously beginning with Marx's comments on drastic new forms of social inequalities and Durkheim's mapping of increasing anomie (or disintegration of social ties) in modern societies, social critics and commentators have continued to theorize *just how badly* modern societies hurt us. By Durkheim's account, the most extreme consequence of anomie was suicide: a social phenomenon which, he pointed out, was strangely most prevalent in those wealthiest segments of society that were least "burdened" by large families and subsequent financial and parental responsibilities. In other words, anomie hit the hardest in segments of society where people were loneliest.

At the dawn of the information age, futurists were imagining the mess that was on its way. In 1960, Alvin Toffler theorized that modern society as a whole was already suffering from "Future Shock," that is to say, extreme confusion from too much change happening over too short a period of time. From our uncertain present at the height of the Facebook age, I propose the phrase "Future Lag" to describe the nonsensical time-space distortions that regulate our existences like a perpetual state of jet lag.

Neal Stephenson's avatars deserve a present-day analogy. But first, a word of clarification. In current scientific lingo, "avatar" refers to the virtual, three-dimensional proxy bodies that are being experimented with by neuroscientists and computer scientists in places like the Stanford Lab for Virtual Humanity (Lanier; Bailenson). Proponents of techno-utopianism like Ray Kurzweil, Jeremy Bailenson, and, to a more cautious extent, Jaron Lanier, have proposed that avatar environments could bring on such miracles as eternal life and the ability to tap into our brain's "pre-adaptive" abilities to do ... well ... *anything*! Jaron Lanier explains that glitches in early avatar experiments (such as finding "oneself" stuck in the hand as opposed to the whole body of an avatar proxy and discovering new possibilities of movement) enabled him and others to experiment with parts of their brainpower that they wouldn't normally use. The argument goes like this: our brains are equipped with the generative ability to use, say, language and writing, just like our bodies are equipped with the possibilities of swimming. But we don't speak, read, or swim unless we learn it socially. We could also configure our life in such ways that we are no longer able to imagine that we could speak, read or write, or swim. What about all the things we can do but cannot imagine?

I lack the space in this chapter to discuss the – in my view, mostly dangerous – avatar experiments currently conducted by Bailenson and others. But Lanier's point about daring to open a plane of possibilities against all that we can no longer imagine is an important one, to which I will return.[2]

---

2 Lanier, who was one of the pioneers of Internet technology, has now mostly turned against the "shallowness" of the current Internet and prefers to build ancient musical instruments.

My argument at this point is that many people *are already living in and addicted to an avatar-sphere*, because they have increasingly come to rely on their semi-fictional online proxies, profiles, and "identities" to interact with each other, themselves, and the world at large. Secondary aspects of my point are easy to grasp: the Internet and social media have provided a hungry void to channel people's fundamental urges to connect with one another in an anomic world but can never provide the real bodily platform to fulfil those desires. The tyranny of speed of our era, furthermore, has turned most of us into progressively more dependent and *desperate* consumers of instant gratification. With the arrival of mobile technologies guaranteeing near-constant access to non-bodily communication across ever decreasing expanses of time and space, we channel our desire for human contact into metaphysical interfaces and find ourselves more *wired* and more *disconnected* at the same time.

The broader ontological questions I introduced earlier (what is real and possible for persons in the world, and what can restrictions of the wired world teach us about it?), however, have yet to be raised. My approach so far has only been object-oriented, or materialistic, in a modestly epistemological sense: I have examined one system of techno-social configurations and identified forms of pathologies (extreme lone-liness, jealousies, bodiless avatars with unfulfillable desires) which, from an ecology of mind perspective (Bateson, *Mind and Nature*; Bateson, *Steps to an Ecology of Mind*), are best understood as pertaining to that system as a whole and not simply to isolated persons; pathology, as Gregory Bateson would have it, is "in the pattern of relationships" (Bateson, *An Ecology of Mind*).

Before returning to the broader possibilities – like river possibilities – that are obfuscated by our current techno-social system, further argu-ments are necessary on why such systems too easily come into being and why they are not unique to "modern," "Western" world views. In theorizing why avatar environments provide a uniquely tempting vortex for "pathological" forms of relationships and personhood, I am inclined to recall another uniquely human compulsion: *our tendency to interact with the world through predominantly symbolic dimensions* – that is to say through reductionistic, primarily metaphysical platforms that "disconnect" us from the world through representational detours. While I lack the space to give an adequate review of literature in the philosophy of mind and related fields that explain these problems with symbolic thought in more detail, suffice it to recall Bateson's definition of metaphysics as an "incorrect" epistemology that prompts us to "confuse the menu card for dinner itself" (Bateson, *Steps to an Ecology of Mind*, 205). An avatar in this sense is merely the representation of a person that comes to be mistaken for, then experienced as, a person herself. Cybernetics (Bateson, *Mind and Nature*; Bateson, *Steps to an Ecology of Mind*; Pickering) and the post-humanities (Wolfe; Latour; Serres) have offered important alternatives on ways to live *adaptively*, or *performatively*, in the world as opposed to *representationally*. But these approaches also tend to rely on simplistic dichotomies between "the west and the rest," where the "Cartesian theatre" (Dennett) that situates personhood in the intangible *cogito* half of Descartes's mind/body model of the human is often seen as an exclusively Western predicament. Post-humanist scholars are particularly unkind to the cognitive sciences, which they see as yet another Western "error" that further entrenches us in representational detours and short-circuitry, since cognitive science scholars are preoccupied with a part of personhood – the mind – that may well be an invention of Western culture. So far so good. Much of this debate lies beyond the scope of this chapter, but my argument on avatars requires that I challenge this last claim and point out that it stems from a rather too simplistic understanding of mind. In Bateson's *An Ecology of Mind*, cognition – or the processing of information – is a property that pertains to a system as a whole. The mind is not so much – or not only – "situated" in the brain as in the entire sensory, bodily, and environmental system in which difference is perceived and enacted. Human epistemologies may require that information be coded and decoded in the brain and translated through cultural specifics, but there would be no such information without bodily, sensory stimulus and a broader material environment. If human eyes see, or a human body runs into, something culturally defined as "a tree," Bateson famously explained, the "mental" work of decoding the information is as immanent in the brain as it is in the eyes, skin, bones ... or the tree. For Bateson, properties of mind – or mental characteristics – in this instance are in the person, the tree, and the close-ended circuit between the person and the tree. And they can indeed travel much farther. We will need to return to the question of how to open and close such "circuits" to non-representational human perception and adaptation. It should suffice to come to the rescue of cognitive scholars, who have clearly shown that most of the work involved in cognition happens at a non-linguistic, intuitive level, is grounded in direct interaction with the environment, and certainly does not require conscious, intentional, explicit symbolic and linguistic translation of information (Bloch). Representational shortcuts, on the other hand, have also been shown to pertain to fundamentally and universally human properties of cognition.

Cognitive anthropologists (Bloch; Boyer; Sperber; Atran) have emphasized *similarities* in human properties of mind and symbolic representations over the cultural particularities and differences that have been the focus in the rest of the discipline. Cognitive anthropology and science of religion in particular has made strong arguments about why and how human properties of mind have given rise to ontologically similar supernatural concepts – or the "presence" of symbolically similar supernatural agents – in all human cultures. I cannot do full justice to these perspectives here[3] and can only summarize some of the key experimental findings in that discipline, which proposes that all humans possess (1) intuitive abilities to project their own subjective and intentional personhoods (or intuitive psychology) onto other agents, "real" or not; (2) an ability to *decouple* reality sensorily into hypothetical situations and detect/project agents accordingly; and (3) structural similarities in the simplified and categorized organization of symbolic information.

In this sense, our ability to feel sorry for an animal, cry at movies, invent imaginary friends, or increase our heart rates when experiencing fear about something that isn't there, hasn't happened, and may never happen all speak to the far reaches of our intersubjectivity with agents and objects that are not human or do not actually exist, or with events that are not actually happening; in other words we are interacting not with the actual (what is the case) but with the potential (what could be the case, may be the case, but isn't). Similar cognitive abilities are at work, Boyer argues, when we create, recall, and make sense of the world through a plethora of supernatural agents (like gods, angels, witches, imaginary friends, magic mountains, etc.) that are assumed to "think" like us.

These highly symbolic detours into the virtual may well be inevitable.

We should now return to a question of scale and possibilities. This takes us to the third point in the cognitive science of religion and an

---

3 See Pascal Boyer's *Religion Explained* for a comprehensive review of the field. One key dimension of the theory is the "epidemiology of culture" (Sperber) framework that proposes a model for how supernatural ideas are propagated, recalled, and found to be constant across time and space in human groups.

exploration of the human propensity for symbolic simplicity. We have seen that online avatars are negotiated through an increasingly mono-cultural consumerist language that is in fact very *simple* and standard-ized. Niche-dependent though it may be (there are different standards on how to be popular, desirable, beautiful, etc.), online identities depend on pre-existing, mostly market-driven patterns of personhood to which subjects attempt to conform. We have also seen that the pro-cess (from the definition of identities to their impossible bodily actual-ization) is *exclusively* symbolic, where online humans have in essence become their own supernatural agents.

I propose to understand this equation through a reversal of common-sense notions about history and complexity: where *increases* in symbolic dependence occurs, ontological simplification ensues; the wired world in this sense is a much narrower, simpler plane of being than any river system of the proverbial savannahs of our hunter-gatherer ancestors.

The notion that the world has become simpler and not more complex is not new to the social sciences. Weber's rationalization and bureauc-ratization thesis provides a good grounding for such an insight. Effi-ciency, Weber argued, was dependent upon functional separation of tasks and people, but also on standardization and efforts to produce predictable, transferable experience. He showed how this came at the cost of quality of personal interaction and brought about alienation. Sociologist George Ritzer, drawing on Weber, articulated his famous McDonaldization thesis to describe post-1950s trends in all aspects of existence (from education and health to leisure and business) to produce schemes that were efficient, predictable, transferable, and extremely simple as structural models. Political scientist James Scott made simi-lar arguments in his excellent review of modernist and high modernist schemes that failed to improve the human condition. Reviewing such cases as the implementation of the metric system, the remodelling of modernist functional cities, and language standardization schemes, Scott discusses centralized impulses to produce controllable *legibility* intent on reducing the "chaos" of the myriad organic ways of being that existed in the world. This came at a considerable human cost.

David Graeber, the anthropologist who became well-known for his involvement in the Occupy Movement, gave us the most monumental work to date to enable us to reflect on the human cost of "progress" as it set out to produce transferable legibility at the expense of long-crafted local ways. His *Debt: The First 5,000 Years* is immensely rich in historical and ethnographic detail and presents the many, many ways that humans had devised to conduct transactions of exchange, through which they constructed and maintained social ties. At the level of what I call official existence, these have all been replaced by the universal, transferable, depersonalized mathematical model of money and capitalism. Graeber makes a compelling case for the violence of this universal system.

Theorists like Michel Serres and Tim Ingold in, respectively, the philosophy and anthropology of knowledge, skill, and writing media have made good arguments about the loss of complexity brought about by technological and symbolic externalizations of knowledge and skill. There is nothing new there either, of course, as Plato was already warning against the effects on memory of the transition from orality to writing.[4] That old debate is worth remembering: what sort of ontologies were possible, we should ask, when, as anthropologist Wade Davis puts it, entire cultural memories were alive at any given moment in the heads of its living members? When not only the *Iliad* or *Mahabharata* could have been committed to memory, but every story, every rule, every skill, every song, every instruction was alive in people's bodies and movements?

Serres and Ingold retrace the history of writing from different perspectives, but both point to the musicality and gesturality that was still alive in the body-praxis of reading and learning until the early modern period and the advent of print media. Reading out loud and sounding out words was the norm and not the exception, they remind us, pointing to medieval engravings of early universities when teachers and scholars seemed to teach and learn standing up, with many hand gestures and amid a puzzling absence of books. Given the rarity of book media and the difficulty of copy in the Middle Ages, Serres explains, scholars had a great amount of textual knowledge committed to memory. What does it entail to affirm that Albertus Magnus synthesized Platonic and Aristotelian knowledge via Arabic translations, when he would have likely never possessed nor been able to copy the textual knowledge for this research? It means he had most or all of it committed to memory!

Art historians have also made significant contributions to the study of continuity between humans and the environment. In *La Vie des Formes*, Henri Focillon in particular took great pains to retrace the history of forms as a translation, or a habitus (my word here) between humans and nature.

---

4 Thoth presenting the invention of writing to King Thamus in the *Phaedrus*.

In his *Éloge de la Main* (Eulogy of the hand), he speculates on the first geometrical and artistic forms, drawn with fingers in the sand of the earth to emulate the lines and twirls of trees, shells, animals, and elements. What kind of disconnection occurred, he asks, when lines in the sand were first drawn with sticks?

Tim Ingold has a lot to say on the subject of lines. As an anthropologist drawing on engagement with hunter-gatherers, a lifetime of research on the perception of the environment, and the intersections between anthropology, archaeology, art, and architecture, he makes fascinating claims about the bodily and spatial discontinuities brought about by technological exteriorizations of knowledge and skill. His small book *Lines: A Brief History* – covering such dimensions as lines of long-form writing, musical notation, lines made in the grass by walking, or lines of airplane travel – identifies the break in continuity between humans and environment at large through a powerful comparison of *navigation* and *wayfaring*. "This distinction between trail-following or wayfaring and pre-planned navigation is of critical significance" (Ingold, *Lines* 15). This fragmentation

> has taken place in the related fields of *travel*, where wayfaring is replaced by destination-oriented transport [or navigation], *mapping*, where the drawn sketch is replaced by the route-plan, and *textuality*, where story-telling is replaced by the pre-composed plot. It has also transformed our understanding of *place*: once a knot tied from multiple and interlaced strands of movement and growth, it now figures as a node in a static network of connectors. (Ingold, *Lines* 75)

Anthropologist Wade Davis invites us into this alternative through the ancient art of Polynesian seafaring:

> [The sea-farer] must process an endless flow of data, intuitions and insights derived from observation and the dynamic rhythms and interactions of wind, waves, clouds, stars, sun, moon, the flight of birds, a bed of kelp, the glow of phosphorescence on a shallow reef – in short, the constantly changing world of weather and the sea. (Davis 60)

But Ingold's analogy between knowledge and travel is telling. The problem of cartography and point-driven navigation is that it is indeed a problem! Or yet another way of confusing dinner with the menu card. My claim here is that symbolic culture, the tunnelled vectors of thought and movement in the official order of things, have conditioned us to move exclusively on and in the map and not in the world it represents.

It often takes the literary imagination of great writers to come to terms with this. Italo Calvino's Marco Polo, for example, warns us that the city should never be confused with the words that describe it; and so all that is left beyond the worlds are … invisible cities. Borges, the great wayfarer of histories and libraries, gives us a wonderful tale on the subject, where the cartographers of Empire had produced a map that was so "exact" that the map itself was exactly the size of Empire and corresponded with it point-by-point. In turn, Georges Perec, that great trickster of language, tells us the joke of space:

> Space begins thus, with words only; signs etched on a blank page. Describing space, naming it, tracing it, like the map-makers who saturated the coasts with names of ports, capes, and names of coves, until the earth, in the end, is only separated from the seas by a ribbon of text. The Aleph, this Borgesian place where the whole world is simultaneously visible, what else is it but an alphabet?
>
> Inventoried space, invented space; space begins with this model map …, 65 geographical terms, miraculously assembled, deliberately abstract: here is the desert, its oasis, its wadi and its chott, here is the source, and there is the stream, the canal, the affluent, the river, the estuary, the mouth and the delta, here is the sea and its islands, its archipelago, its inlets, its reefs, its waves, its surf, its coastal line, here is the strait, the isthmus, the peninsula, the gulf and the bay, the cape and the lagoon, the cliff, the beach, the pond, the lake and the marsh, and here are the mountains, the glacier, the volcano, the buttress, the range, the ridge, the flank, the plateau, the hill; here is the city and its harbour, its port, its lighthouse. (Perec 4, my translation)

Conclusion: Back to the Jungle

This chapter has presented a long list of culprits in the invention, projection, inventorization, standardization, and regulation of an increasingly narrow and simple world, with human relationships on the Internet as a case in point. I have argued that the wired world and its avatars, despite promises of speed, connections, and complexity, offers the most aggressive prison devised so far in restricting the terms of human personhood and relations with each other and the world.

Navigating through various theories of the human from relativist to universalist positions, I presented a cautious argument on a generalized set of human problems (symbolic thought, language, symbol-dependent technologies) against possibilities of relationships where humans are at once more metaphysically fluid and materially grounded (multi-naturalism, wayfaring).

The opening stories of my own tentative wayfaring in the Amazon offered a glimpse into my own proto-theoretical journey in identify-ing problems of avatar personhoods in the wired world, while at the same time inviting me to engage with an open-ended plane of realities, relationships and possibilities. The "jungle," as I have come to define it since, beyond its slow and capillary rivers and hybrid persons that eluded international borders and most institutionalized forms of culture, offered by definition a space *that is not finite*. Like the motorboat pilot who "knew" every bend, rock, and moving current in the black river, the jungle requires a *performative*, constantly re-adaptive epistemology that no representational system can contain; it requires, in broader terms, what Pickering calls an "ontology of unknowability," a world that is at once slower and harder, but also faster and softer, fluid and fixed, and infinitely more complex than any modern techno-social system. As an extreme in that last genre, the wired world has little to offer besides its own unfulfillable game. That kind of "ontological monotheism," as Pick-ering remarked, "is not turning out to be a pretty sight" (33).

For a prettier sight – and infinite experience – I suggest you find a river.

WORKS CITED

Atran, Scott. *In Gods We Trust: The Evolutionary Landscape of Religion*. Oxford: Oxford University Press, 2002. Print.

Badiou, Alain, Graham Harman, and Quentin Meillassoux. *Realizm Spekulatywny: Przygodność, Wirtualność, Konieczność*. Warsaw: Fundacja Augusta Hr. Cieszkowskiego, 2012. Print.

Bailenson, Jeremy. *Infinite Reality: Avatars, Eternal Life, New Worlds, and the Dawn of the Virtual Revolution*. New York: William Morrow, 2011. Print.

Bateson, Gregory. *Mind and Nature: A Necessary Unity*. New York: Dutton, 1979. Print.

– *Steps to an Ecology of Mind: Collected Essays in Anthropology, Psychiatry, Evolution, and Epistemology*. San Francisco: Chandler Pub., 1972. Print.

Bateson, Nora, dir. *An Ecology of Mind*. Oley, PA: Bullfrog Films, 2011.

Bloch, Maurice. *How We Think They Think: Anthropological Approaches to Cognition, Memory, and Literacy*. Boulder, CO: Westview, 1998. Print.

Borges, Jorge Luis. *Collected Fictions*. Trans. Andrew Hurley. New York: Viking, 1998. Print.

Boyer, Pascal. *Religion Explained: The Evolutionary Origins of Religious Thought*. New York: Basic, 2001. Print.

Bryant, Levi R., Nick Srnicek, and Graham Harman. *The Speculative Turn: Continental Materialism and Realism*. Melbourne: Re.press, 2011. Print.

Calvino, Italo. *Invisible Cities*. New York: Harcourt Brace Jovanovich, 1974. Print.

Davis, Wade. *The Wayfinders: Why Ancient Wisdom Matters in the Modern World*. Toronto: House of Anansi, 2009. Print.

Dennett, D.C. *Consciousness Explained*. Boston: Little, Brown and Co., 1991. Print.

Derrida, Jacques. *Writing and Difference*. Chicago: University of Chicago Press, 1978. Print.

Descola, Philippe. *In the Society of Nature: A Native Ecology in Amazonia*. Cambridge: Cambridge University Press, 1994. Print.

– *The Spears of Twilight: Life and Death in the Amazon Jungle*. New York: New Press, 1996. Print.

Durkheim, Émile. *Suicide, a Study in Sociology*. Glencoe, IL: Free Press, 1951. Print.

Focillon, Henri. *Éloge de la Main*. Paris: Presses Universitaires De France, 1934. Print.

– *La Vie des Formes*. Paris: Presses Universitaires De France, 1947. Print.

Graeber, David. *Debt: The First 5,000 Years*. Brooklyn: Melville House, 2011. Print.

Grenoble, Ryan. "Amanda Todd: Bullied Canadian Teen Commits Suicide after Prolonged Battle Online and in School." *Huffington Post*. TheHuffingtonPost.com, 11 Oct. 2012. Web. 26 Feb. 2014.

Harman, Graham. "Technology, Objects and Things in Heidegger." *Cambridge Journal of Economics* 34.1 (2010): 17–25. Print.

– *Tool-being: Heidegger and the Metaphyics of Objects*. Chicago: Open Court, 2002. Print.

Heidegger, Martin. *Bremer Und Freiburger Vorträge*. Frankfurt Am Main: Klostermann, 1994. Print.

– *The Question Concerning Technology, and Other Essays*. New York: Harper & Row, 1977. Print.

Heidegger, Martin, W.B. Barton, and Vera Deutsch. *What Is a Thing?* Chicago: H. Regnery, 1968. Print.

Herzog, Werner. "Herzog on the Obscenity of the Jungle." *YouTube*. YouTube, 12 June 2006. Web. 1 Mar. 2014.

Ingold, Tim. *Being Alive: Essays on Movement, Knowledge and Description*. London: Routledge, 2011. Print.

– *Lines: A Brief History*. London: Routledge, 2007. Print.

– *The Perception of the Environment: Essays on Livelihood, Dwelling & Skill*. London: Routledge, 2000. Print.

Keane, Webb. "Marked, Absent, Habitual: Approaches to Neolithic Religion at Çatalhöyük." *Religion in the Emergence of Civilization: Çatalhöyük As a Case Study*. Ed. Ian Hodder. Cambridge: Cambridge University Press, 2010. 187–219. Print.

Kehlmann, Daniel, and Carol Brown Janeway. *Measuring the World*. New York: Pantheon, 2006. Print.

Kohn, Eduardo. *How Forests Think: Toward an Anthropology beyond the Human*. Berkeley: University of California, 2013. Print.

Kurzweil, Ray. *The Singularity Is Near: When Humans Transcend Biology*. New York: Viking, 2005. Print.

Lanier, Jaron. *You Are Not a Gadget: A Manifesto*. New York: Alfred A. Knopf, 2010. Print.

Latour, Bruno. *Politics of Nature: How to Bring the Sciences into Democracy*. Cambridge, MA: Harvard University Press, 2004. Print.

Marx, Karl. *Capital, Volume 1: A Critique of Political Economy*. New York: International Publishers, 1977. Print.

Meillassoux, Quentin. *After Finitude: An Essay on the Necessity of Contingency*. London: Continuum, 2008. Print.

Olsen, Bjørnar. *In Defense of Things: Archaeology and the Ontology of Objects*. Lanham, MD: AltaMira Press, 2010. Print.

Parikka, Jussi. *Insect Media: An Archaeology of Animals and Technology*. Minneapolis: University of Minnesota, 2010. Print.

Perec, Georges. *Espèces D'espaces*. 32, Rue Du Fer-à-Moulin [Paris]: Editions Galilée, 1974. Print.

Pickering, Andrew. *The Cybernetic Brain: Sketches of Another Future*. Chicago: University of Chicago Press, 2010. Print.

Plato. *Phaedrus*. Trans. C.J. Rowe. Warminster, Wiltshire, UK: Aris & Phillips, 1988. Print.

Ritzer, George. *The McDonaldization of Society*. Thousand Oaks, CA: Pine Forge, 2004. Print.

Scott, James C. *Seeing Like a State: How Certain Schemes to Improve the Human Condition Have Failed*. New Haven, CT: Yale University Press, 1998. Print.

Serres, Michel. *Petite Poucette*. Paris: Éd. Le Pommier, 2012. Print.

Sperber, Dan. *Explaining Culture: A Naturalistic Approach*. Oxford: Blackwell, 1996. Print.

Stengers, Isabelle. *Cosmopolitics I*. Minneapolis: University of Minnesota Press, 2010. Print.

Stephenson, Neal. *Snow Crash*. New York: Bantam, 1992. Print.

Sterelny, Kim. *The Evolved Apprentice: How Evolution Made Humans Unique.* Cambridge, MA: MIT Press, 2012. Print.

Toffler, Alvin. *Future Shock*. New York: Random House, 1970. Print.

Tomasello, Michael. *A Natural History of Human Thinking*. Cambridge, MA: Harvard University Press, 2014. Print.

Viveiros de Castro, Eduardo Batalha. *A Inconstância da Alma Selvagem e Outros Ensaios de Antropologia*. São Paulo: Cosac & Naify, 2002. Print.

– "Cosmological Deixis and Amerindian Perspectivism." *Journal of the Royal Anthropological Institute* 4.3 (1998): 469–88. Print.

– "Perspectival Anthropology and the Method of Controlled Equivocation." *Tipití: Journal of the Society for the Anthropology of Lowland South America* 2.1 (2004): 1–20. Print.

Wagner, Roy. *An Anthropology of the Subject: Holographic Worldview in New Guinea and Its Meaning and Significance for the World of Anthropology.* Berkeley: University of California Press, 2001. Print.

Weber, Max. *Max Weber on Capitalism, Bureaucracy, and Religion: A Selection of Texts*. Comp. Stanislav Andreski. London: Allen & Unwin, 1983. Print.

Wolfe, Cary. *What Is Posthumanism?* Minneapolis: University of Minnesota Press, 2010. Print.

Contributors

Ahmet Atay is an assistant professor at the College of Wooster. He has an M.A degree from Ohio University (Telecommunications), an MA degree from the University of Northern Iowa (Communication Studies), and a PhD from Southern Illinois University, Carbondale in Intercultural/International Communication, Postcolonial Studies, and Media Studies. His research interests include representation of cultural identity in cyberspace, diasporic experiences, representation of gender and sexuality in soap operas, Italian cinema and queer identities, and audience studies. In particular, he has been studying diasporic identity formations of queer bodies in cyberspace through cyber ethnography.

Aparajita De is an assistant professor in the Department of Geography, Delhi School of Economics, University of Delhi. Her research interests include media and popular culture.

Laura E. Enriquez is an assistant professor in the Department of Chicano and Latino Studies at the University of California, Irvine. She received her PhD in Sociology from the University of California, Los Angeles. She specializes in the study of immigration, race/ethnicity, and citizenship. Her recent work focuses on the educational, political, and social incorporation experiences of undocumented immigrant young adults. She is currently working on a book manuscript that examines how immigration laws shape the family formation experiences of undocumented young adults in the United States.

May Friedman blends social work, teaching, research, writing, and parenting. May's passions include social justice and reality TV (she is firmly in favour of living with contradiction). Recent publications include work on motherhood and transnationalism, gender fluidity, and *Here Comes Honey Boo Boo*. A faculty member in the Ryerson University School of Social Work and Ryerson/York graduate program in Communication and Culture, May lives in downtown Toronto with her partner and four young children.

Kimberly McKee is an assistant professor in the Department of Liberal Studies at Grand Valley State University. She received her PhD in Women's, Gender, and Sexuality Studies from the Ohio State University. She is currently revising her book manuscript that examines the institutional practice of international adoption, traces the origins of what she terms the transnational adoption industrial complex – the neocolonial, multi-million-dollar industry that commodifies children's bodies and centres the voices of adult adoptees.

Shekh Moinuddin currently teaches at Kalindi College, University of Delhi and pursues a PhD in the Department of Geography, Delhi School of Economics, University of Delhi.

Isabella Ng is an assistant professor in the Hong Kong Institute of Education's Department of Asian and Policy Studies. She received her PhD in Gender Studies from the School of Oriental and African Studies (SOAS). Her research interests include gender and development in Asia Pacific, anthropological study of rural villages in Hong Kong and China, and transnational migration and anthropology of media.

Julia Obermayr received a Master of Arts in Spanish and French, studied Austrian Sign Language, and taught at both the Departments for Romance and for Translation Studies at the University of Graz, Austria. In her doctoral thesis she is currently analysing representations of lesbians in Web series and their effect on audiences.

Silvia Schultermandl is an assistant professor of American Studies at the University of Graz, Austria, specializing in transnational American studies. She is currently at work on a monograph on the aesthetics of transnationalism in American literature from the Revolution to 9/11. Together with Erin Kenny (Drury University), she is the series editor of LIT Verlag's book series Contributions to Transnational Feminism.

Samuel Veissière is an assistant professor of Anthropology at the University College of the North and holds a visiting professorship in the Culture, Mind, and Brain program at McGill University, where he teaches in the Division of Social and Transcultural Psychiatry and the Department of Anthropology. He has conducted fieldwork on emergent modes of sociality and cooperation in the context of street children's livelihoods and transnational sex work in Brazil, "clandestine" migration in the Pan-Amazon, Indigenous revivalism, and various new forms of Internet cultures. His current work investigates the mediation of "Tulpa" sentient imaginary friends on the Internet and ways of being a child and learning about "persons" across cultures.

M. Tina Zarpour earned her Master of Applied Anthropology and PhD in Socio-Cultural Anthropology from the University of Maryland, College Park. She continues her research on local immigrant communities and their intersections with civic and political life as an independent scholar while also teaching, and she serves as the Director of Education at the San Diego History Center. Her recent work is about the incorporation of sub-altern heritages (through narratives from recent migrants) into what is considered the official history of the region.

Index

Lightning Source UK Ltd.
Milton Keynes UK
UKOW01f2201100516

273973UK00001B/80/P